高校专门用途英语(ESP)系列规划教材

航空基础英语
Aviation English Basics

主　编　刘　泉　王孝存　王建颖
副主编　高　峰　郭家宁　阿　荣
编　者　白旭阳　马国磊　张占伟
　　　　凌　晨　王　鑫

北京师范大学出版集团
BEIJING NORMAL UNIVERSITY PUBLISHING GROUP
安徽大学出版社

图书在版编目（CIP）数据

航空基础英语 / 刘泉，王孝存，王建颖主编 . -- 合肥：安徽大学出版社，2024.9

高校专门用途英语（ESP）系列规划教材

ISBN 978-7-5664-2798-4

Ⅰ . ①航… Ⅱ . ①刘…②王…③王… Ⅲ . ①航空工程—英语—高等学校—教材 Ⅳ . ① V2

中国国家版本馆 CIP 数据核字（2024）第 021546 号

航空基础英语
Hangkong Jichu Yingyu

刘　泉　王孝存　王建颖　主编

出版发行：	北京师范大学出版集团 安徽大学出版社 （安徽省合肥市肥西路 3 号 邮编 230039） www.bnupg.com www.ahupress.com.cn
印　　刷：	安徽利民印务有限公司
经　　销：	全国新华书店
开　　本：	787 mm × 1092 mm　1/16
印　　张：	14.5
字　　数：	434 千字
版　　次：	2024 年 9 月第 1 版
印　　次：	2024 年 9 月第 1 次印刷
定　　价：	49.90 元

ISBN 978-7-5664-2798-4

策划编辑：李　雪		装帧设计：李　军	
责任编辑：李　雪		美术编辑：李　军	
责任校对：高婷婷		责任印制：陈　如　孟献辉	

版权所有　侵权必究

反盗版、侵权举报电话：0551-65106311
外埠邮购电话：0551-65107716
本书如有印装质量问题，请与印制管理部联系调换。
印制管理部电话：0551-65106311

Preface
前 言

　　教育部办公厅《"十四五"普通高等教育本科国家级规划教材建设实施方案》推动了"产教融合、科教融汇"的协同育人培养模式改革，鼓励跨校、跨区域联合编写教材。《大学英语教学指南（2020版）》要求：各高校应以需求分析为基础，根据学校人才培养规格和学生需要开设体现学校特色的专门用途英语课程；也可在通用英语体系内，纳入通用学术英语和职业英语等内容。"十四五"规划纲要指出：要深化职普融通、产教融合、校企合作。

　　编者按照以上文件精神，根据民航局最新规定和民航业的切实需求，针对目前的航空专业英语教材内容与形式单一、通识内容与专业内容缺乏融合、学生实际语言运用能力和文化意识培养不足、配套教辅资料不够、教师备课工作量大等问题，以及航空专业学生就业时同时持有航空专业英语证书和大学英语四级证书的需求，编写了这本衔接通识英语与航空专业英语的教材，旨在帮助学生通过相关考试、培养职业技能，为其今后的职业发展打下基础。

　　本教材共12个单元，阅读文本长度为800词至1200词，生词率为7%左右。以航空话题为单元主线，结合了基础英语的培养。每单元共有7个版块。

　　第1版块为单元预览，围绕单元主题设置导语，用平实易懂的语言对本单元内容进行全面介绍，建立学习者对本单元话题的整体认识。

　　第2版块为听力，围绕单元主题设计听力练习，引起学习者对相关话题的探究兴趣。

　　第3版块为阅读，是本教材的重点板块，包括本单元课文、生词、文章体裁分析、文化小贴士和3项相关练习。在生词部分中，黑体的词为近年四级考试中出现的词汇；标▲的词为重点词汇。文章体裁分析包括体裁辨识、交际语步辨识和交际目的分析。文化小贴士提供了跨文化交际方面的建议。

第4版块为翻译，围绕单元相关语言的功能性结构及核心要点设置翻译练习，培养学习者的文本解读、转述以及实际应用能力。

第5版块为口语，材料均为真实应用场景中行业相关人员的典型职场交际口语，培养学生的口头表达能力。

第6版块为写作。

第7版块提供了相关主题的四级真题拓展练习。

本教材具有以下特色：

- 融合航空英语与基础英语内容，为航空专业英语学习打下坚实的基础。
- 配套多模态音视频学习资料。鄂尔多斯应用技术学院自建的微信小程序"云梯YunTeach"提供本教材配套的核心词汇课程，内容包括各单元单词和句子的朗读和口译，并具有测评功能。
- 课程思政元素突出，其中航空人文单元介绍了我国"航空之父"冯如；航空趋势单元的内容为C919国产大飞机的研发历程。

本教材是由来自三所院校（鄂尔多斯应用技术学院、南京航空航天大学、内蒙古工业大学）的教师、中国商用飞机有限公司高级培训讲师和工程师，以及优秀大学生组成的团队编写的，充分体现了产教融合和跨校、跨区域联合编写教材的理念。教材编写受到了鄂尔多斯应用技术学院外籍教师David Casole的大力支持。

本教材能够帮助学生形成对航空业的宏观认识，习得跨文化知识和跨学科知识，提升英语综合能力。其可作为航空专业学生教材，航空公司、维修单位和机场的培训教材，以及大众科普读物。

编　者
2024年9月

★ CONTENTS ★

Unit 1 Aviation Health 1
    Preview 1
    Listening 2
    Reading 2
    Translation 14
    Speaking 15
    Writing 16
    Expansion Exercise for CET-BAND 4—Reading 16

Unit 2 Aviation Science 19
    Preview 19
    Listening 20
    Reading 20
    Translation 32
    Speaking 33
    Writing 34
    Expansion Exercise for CET-BAND 4—Reading 35

Unit 3 Aviation Humanity 37
    Preview 37
    Listening 38
    Reading 38

Translation …………………………………………………………… 50
Speaking …………………………………………………………… 51
Writing ……………………………………………………………… 52
Expansion Exercise for CET-BAND 4—Reading ………………… 53

Unit 4 Aviation Education …………………………………………… 55
Preview ……………………………………………………………… 55
Listening …………………………………………………………… 56
Reading …………………………………………………………… 56
Translation ………………………………………………………… 67
Speaking …………………………………………………………… 69
Writing ……………………………………………………………… 70
Expansion Exercise for CET-BAND 4—Reading ………………… 71

Unit 5 Aviation Legend ……………………………………………… 73
Preview ……………………………………………………………… 73
Listening …………………………………………………………… 74
Reading …………………………………………………………… 74
Translation ………………………………………………………… 83
Speaking …………………………………………………………… 84
Writing ……………………………………………………………… 85
Expansion Exercise for CET-BAND 4—Reading ………………… 85

Unit 6 Aviation Feat ………………………………………………… 87
Preview ……………………………………………………………… 87
Listening …………………………………………………………… 88
Reading …………………………………………………………… 89
Translation ………………………………………………………… 96
Speaking …………………………………………………………… 98
Writing ……………………………………………………………… 99

Expansion Exercise for CET-BAND 4—News Report 100

Unit 7 Aviation Technology 101
    Preview 101
    Listening 102
    Reading 102
    Translation 113
    Speaking 114
    Writing 114
    Expansion Exercise for CET-BAND 4—News Report 115

Unit 8 Aviation Anecdote 117
    Preview 117
    Listening 118
    Reading 118
    Translation 128
    Speaking 128
    Writing 129
    Expansion Exercise for CET-BAND 4—Reading 130

Unit 9 Aviation Economy 135
    Preview 135
    Listening 136
    Reading 136
    Translation 145
    Speaking 146
    Writing 148
    Expansion Exercise for CET-BAND 4—Reading 148

Unit 10 Aviation Trend 151
Preview 151
Listening 152
Reading 152
Translation 161
Speaking 162
Writing 163
Expansion Exercise for CET-BAND 4—Translation 164

Unit 11 Aviation Energy 165
Preview 165
Listening 166
Reading 166
Translation 178
Speaking 179
Writing 180
Expansion Exercise for CET-BAND 4—News Report 180

Unit 12 Aviation Accident 183
Preview 183
Listening 184
Reading 185
Translation 198
Speaking 199
Writing 200
Expansion Exercise for CET-BAND 4—News Report 201

Keys 203
References 220

Unit 1　Aviation Health

Preview

　　The passage discusses the development of digital health passports, such as AOK-pass, as a means to verify COVID-19 test results or vaccination status. These passports aim to replace paper immunization cards and are being tested by airports and airlines worldwide. The process involves downloading an app, inputting personal information, and obtaining a unique QR code as proof of a negative test result. Concerns over data privacy and potential leaks exist, but efforts are made to encrypt and minimize data collection. While not widely implemented, the passports are seen as a game changer for post-pandemic travel, although their adoption varies among airlines.

Listening

Directions: *Fill in the blanks with the words that you hear. Find them in the table below and write down the corresponding letters.*

AOK-pass, a digital health passport that __1__ whether someone has tested negative for the coronavirus or gotten inoculated. It's also known as a vaccine passport and meant to be easier to use and more __2__ than paper immunization cards. Travel industry experts say this type of digital health certificate could be a game changer, not just for tourism, but live events and schools. So here's how digital health passports __3__.

Passengers can __4__ the AOK-pass app and input their personal information like their birthday and email address. Within 30 minutes, the result is __5__. If the traveler tests negative, the app creates a unique QR code that serves as proof. Some but not all airlines and governments require proof of a negative test before a passenger is __6__ to board a flight or enter a country. While testing and vaccinations reduce the risk of __7__ the virus, they don't eliminate it.

"In the setting of COVID-19, health information is beginning to be used in a transitional sense, much more than before." Chester Drum is the co-founder of AOK-pass. He expects that in order for people to __8__ to pre-pandemic life, they'll have to show proof of their health status. Public health researchers have found that some people in India and Indonesia who tested positive and have their results disclosed were verbally __9__ or socially excluded. Drum says figuring out the technical parts of the digital health passport were __10__ easy, compared with making sure the app would be used along each step of the travel process.

| A. spreading | B. reliable | C. relatively | D. available | E. return |
| F. work | G. verifies | H. attacked | I. allowed | J. download |

Reading

How a Digital Health Passport Could Work

Giulio Castagnara is preparing to fly to New York, but before he can **board** his flight, he must provide proof that he is **COVID**-free. Fortunately, there is a convenient solution called AOK-pass, a **digital** health passport that **verifies** a person's **negative** COVID-19 test results or **vaccination status**. Unlike traditional paper **immunization** cards, AOK-pass is an **app** that can be **download**ed onto a smartphone and contains all the necessary information. It aims to simplify the process and increase reliability.

The use of digital health passports has gained momentum in response to the **devastating** impact of the **pandemic** on the travel industry. Airports in Rome, Paris, and Singapore, along with airlines like JetBlue and United, are experimenting with similar apps. They are all **vying** for a share of the global market for travel health services and **certification**s, **estimated** to be worth nearly $20 billion. Experts believe that these digital health certificates have the potential to be a game changer not only for **tourism** but also for live events and schools.

So, how does a digital health passport like AOK-pass actually work? Let's take a closer look. At airports like the one in Rome, travelers who were unable to take a COVID-19 test before **departure** can undergo testing at a designated site **set up** by the airport. While waiting for their test results, passengers can download the AOK-pass app and **input** their personal information, such as their birthday and email address. Within a short period, usually 30 minutes, the test results become **available**. If the traveler tests negative, the app generates a unique QR code that serves as **proof** of their health status.

The QR code plays a crucial role in the process. At the airport, it is **scan**ned during check-in and boarding. By scanning the code, airline staff or medical professionals can quickly verify that the traveler has tested negative for COVID-19. This verification process ensures safer travel by reducing **the risk of** spreading the virus. While some airlines and governments **mandate** proof of a negative test before allowing passengers to board or enter a country, it's important to note that testing and vaccinations reduce the risk of **transmission** but do not **eliminate** it entirely.

Chester Drum, a doctor at the National University Hospital in Singapore and co-founder of AOK-pass, believes that **showcasing** proof of health status will become a common requirement for people to return to pre-pandemic activities. **For instance**, individuals may need to present evidence of vaccination to attend events or display their medical records to board an airplane. Digital health passports like AOK-pass aim to **facilitate** these processes efficiently and securely.

One significant advantage of using **blockchain** technology in these apps is the prevention of **forgeries**. In the past, incidents of fake COVID-19 test results being sold at high prices have occurred. However, AOK-pass **tackle**s this issue by **utilizing** blockchain, ensuring that each test result and vaccination certificate is unique and **linked to** a specific **individual**. Once data is entered into the blockchain, it becomes **tamper**-proof, **enhancing** security and trust.

Nevertheless, there are concerns regarding data **privacy** and **leaks**. Ana Badeuschi, a researcher studying the **ethical implications** of digital health data, emphasizes the need for policymakers to require digital health passport providers to adopt appropriate measures to protect data privacy and **address** potential  **invasive** situations. Public health researchers have found instances of people being **verbally attacked** or socially **excluded** after their positive test results were disclosed. To **mitigate** such risks, maintaining the **confidentiality** of test results is recommended.

AOK-pass takes privacy seriously and **encrypts** all personal health information. The company also claims not to **monetize** user data. It has signed **contract**s with airports and airlines to develop and test the system. After nearly ten months of testing, AOK-pass is **schedule**d to **launch** in major Italian and French airports by the end of February. However, it will face competition from other companies developing similar tools. For example, United and JetBlue plan to use an app developed by the World Economic Forum called the Common Pass, while IBM is working on a digital health pass using blockchain technology.

Chester Drum **acknowledges** that the technical aspects of the digital health passport were relatively easy to **figure out** compared to ensuring its **seamless integration** throughout the entire travel process. Achieving **coordination** among different **entities**, such as laboratories, airlines, and immigration authorities, **poses** significant challenges due to the varying regulations and **procedures** involved.

When Giulio Castagnara arrived in New York during a layover on his way to Dallas, he noticed that the airline staff did not ask for proof of his health status. Although the **implementation** of the app may vary, Castagnara still considers the digital health passport a game changer. Having last traveled to the States in March 2020, just before the pandemic struck, he **appreciate**s the ability to travel again with the added **assurance** of safety provided by the digital health passport.

New Words & Expressions

board *v.* /bɔː(r)d/		to get on a ship, train, plane, bus, etc.
		上船或火车、飞机、公共汽车等

Unit 1 Aviation Health

COVID *n.* /ˈkəʊvɪd/	Corona Virus Disease 新型冠状病毒肺炎
coronavirus *n.* /kəˈrəʊnəvaɪrəs/	any of a group of RNA viruses that cause a variety of diseases in humans and other animals 冠状病毒，冠形病毒
▲**digital** *adj.* /ˈdɪdʒɪt(ə)l/	of a circuit or device that represents magnitudes in digits 数字信息系统的，数码的，数字式的，数位的
verify *v.* /ˈverɪfaɪ/	to check or to prove that something is true or correct 证实，证明，核验
negative *n.* /ˈneɡətɪv/	not showing any evidence of a particular substance or medical condition 结果为阴性的（或否定的）
vaccination *n.* /ˌvæksɪˈneɪʃ(ə)n/	taking a vaccine as a precaution against contracting a disease ［医］预防接种，疫苗接种，种痘
status *n.* /ˈsteɪtəs/	the social or professional position of sb./sth. in relation to others; the situation at a particular time during a process 地位，身份，（进展的）状况，情形
immunization *n.* /ˌɪmjʊnaɪˈzeɪʃn/	the act of making sth. secure against infections or diseases (esp. by inoculation) 免疫，免疫作用，免疫法，免除
app / abbr. /æp/	the abbreviation for "application", a program designed to do a particular job; a piece of software 应用程序，应用软件
download *v.* /ˌdaʊnˈləʊd/	to move data to a computer system from a central one 下载
devastate *v.* /ˈdevəsteɪt/	to completely destroy a place or an area 彻底破坏，摧毁，毁灭

pandemic *n.* /pænˈdemɪk/	a disease that spreads over a whole country or the whole world （全国或全球性）流行病，大流行病
vie *v.* /vaɪ/	(vying, vied, vied) to compete strongly with sb. in order to obtain or achieve sth. 激烈竞争，争夺
certification *n.* /ˌsɜ:tɪfɪˈkeɪʃn/	the act to provide evidence for sth. 证明，鉴定
▲estimate *v.* /ˈestɪmeɪt/	a judgement that you make without having the exact details of sth. 估计，预算，预测，评价
tourism *n.* /ˈtʊərɪz(ə)m/	the business activity connected with providing accommodation, services and entertainment for people who are visiting a place for pleasure 旅游业，观光业
▲departure *v.* /dɪˈpɑ:(r)tʃə(r)/	the act of leaving a place; an example of this 离开，起程，出发
input *v.* /ˈɪnpʊt/	time, knowledge, ideas, etc. that you put into work, a project, etc. 投入资源（指时间、知识、思想等），投入，输入
available *adj.* /əˈveɪləb(ə)l/	that you can get, buy or find 可获得的，可购得的，可找到的
proof *n.* /pru:f/	information, documents, etc. showing that sth. is true 证据，证明
scan *v.* /skæn/	to look quickly but not very carefully at a document, etc.; (of a light, radar, etc.) to pass across an area 粗略地读，浏览，翻阅，（光束、雷达等）扫描，扫掠
mandate *v.* /ˈmændeɪt/	to order sb. to do sth. or vote in a particular way 强制执行，委托办理

transmission *n.* /træns'mɪʃn/	the act or process of passing sth. from one person, place or thing to another
	传送，传递，传达，传播，传染
eliminate *v.* /ɪ'lɪmɪˌneɪt/	to remove or get rid of sth./sb.
	排除，清除，消除
showcase *v.* /'ʃəʊkeɪs/	exhibit, display
	展示，表现
facilitate *v.* /fə'sɪlɪteɪt/	to make an action or a process possible or easier
	促进，促使，使便利
blockchain *n.* /'blɒkˌtʃeɪn/	a distributed accounting journal with growing lists of records (blocks) that are linked together
	车链，区块链
forgery *n.* /'fɔː(r)dʒəri/	fake money or product
	伪造品，赝品
tackle *v.* /'tæk(ə)l/	accept as a challenge
	处理
utilize *v.* /'juːtəlaɪz/	to use sth., especially for a practical purpose
	使用，利用，运用，应用
▲individual *adj.* /ˌɪndɪ'vɪdʒuəl/	considered separately rather than as part of a group
	单独的，个别的
n.	a person considered separately rather than as part of a group
	个人
tamper *v.* /'tæmpə(r)/	to make changes to sth. without permission, especially in order to damage it
	篡改，擅自改动，胡乱摆弄（尤指有意破坏）
enhance *v.* /ɪn'hɑːns/	to increase or further improve the good quality, value or status of sb./sth.
	提高，增强，增进
▲privacy *n.* /'praɪvəsi/	the state of being alone, not watched or disturbed by other people; being free from

	public attention
	隐私，私密
leak *n.* /liːk/	a deliberate act of giving secret information to sb.
	（秘密信息的）透露
ethical *adj.* /ˈeθɪk(ə)l/	connected with beliefs and principles about what is right and wrong
	（有关）道德的，伦理的
implication *n.* /ˌɪmplɪˈkeɪʃ(ə)n/	a possible effect or result of an action or a decision
	可能的影响（或作用、结果）
address *v.* /əˈdres/	to think about a problem or a situation and decide how you are going to deal with it
	设法解决，处理，对付
invasive *adj.* /ɪnˈveɪsɪv/	entering and spreading very quickly and difficult to stop
	侵入的，侵袭的
verbally *adv.* /ˈvɜː(r)bəli/	in spoken words and not in writing or actions
	口头上
attack *v.* /əˈtæk/	an act of using violence to try to hurt or kill sb./sth.
	袭击，攻击
▲exclude *v.* /ɪkˈskluːd/	prevent sth. from being included or considered
	不包括，不放在考虑之列
mitigate *v.* /ˈmɪtɪgeɪt/	to make sth. less harmful, serious, etc.
	减轻，缓和
confidentiality *n.* /ˌkɒnfɪˌdenʃɪˈæləti/	a situation in which you expect sb. to keep information secret
	保密性，机密性
encrypt *v.* /ɪnˈkrɪpt/	to put information into a special code
	把……加密（或编码）
monetize *v.* /ˈmʌnɪtaɪz/	convert into or express in the form of

	currency
	把……作为法定货币，使具有货币性质，把……铸成货币
contract *n.* /ˈkɒntrækt/	an official written agreement
	合同，合约，契约
v.	to make a pact with sb.
	签合同，订立契约
▲schedule *v.* /ˈʃedjuːl/ /ˈskedʒuːl/	to arrange for sth. to happen at a particular time
	安排，为……安排时间，预定
▲launch *v.* /lɔːntʃ/	to make a product available to the public for the first time
	（首次）上市，发行
acknowledge *v.* /əkˈnɒlɪdʒ/	to accept that sth. is true
	承认（属实）
seamless *adj.* /ˈsiːmləs/	with no spaces or pauses between one part and the next
	（两部分之间）无空隙的，不停顿的
integration *n.* /ˌɪntɪˈɡreɪʃn/	the act or process of combining two or more things so that they work together
	结合，整合，一体化
coordination *n.* /kəʊˌɔːdɪˈneɪʃn/	the act of making parts of sth., groups of people, etc. work together in an efficient and organized way
	协作，协调，配合
entity *n.* /ˈentəti/	something that exists separately from other things and has its own identity
	独立存在物，实体
pose *v.* /pəʊz/	to create a threat, problem, etc. that has to be dealt with
	造成（威胁、问题等），引起，产生
procedure *n.* /prəˈsiːdʒə(r)/	a way of doing sth., especially the usual or correct way
	程序，手续，步骤

implementation *n.* /ˌɪmplɪmenˈteɪʃ(ə)/	the act of accomplishing some aim or executing some order
	实施，执行
appreciate *v.* /əˈpriːʃieɪt/	to recognize the good qualities of sb./sth.
	欣赏，赏识，重视
assurance *n.* /əˈʃʊərəns/	a statement that sth. will certainly be true or will certainly happen
	保证，担保
set up	建立，创立，发起，开办
the risk of ...	……的危险
in order for / to	为了
for instance	例如
link to	链接
figure out	估计，解决，了解，[口]确定

Genre Analysis

Genre:

The above passage can be classified as an informative article or news piece discussing the concept and implementation of digital health passports for travel.

Communicative Moves:

Move 1 Providing Background Information:

The passage begins by introducing Giulio Castagnara and his need to provide the proof of being COVID-free before traveling. It introduces the concept of a digital health passport called AOK-pass and highlights its convenience and reliability compared to traditional paper immunization cards.

Move 2 Presenting the Global Trend:

This move discusses the increasing popularity of digital health passports in response to the pandemic's impact on the travel industry. It mentions various airports and airlines experimenting with similar apps and emphasizes the potential market value of travel health services and certifications.

Move 3 Explaining the Process:

The passage delves into the actual workings of a digital health passport like AOK-pass. It describes how travelers can undergo testing at designated sites, download the app, and input personal information. It explains that a negative test generates a unique QR code serving as

Unit 1 Aviation Health

proof of health status.

Move 4 Highlighting the Role of QR Codes:

This move focuses on the importance of QR codes in the verification process during check-in and boarding. It highlights the significance of scanning the code to ensure safer travel and mentions that testing and vaccinations reduce but do not eliminate the risk of transmission.

Move 5 Discussing Future Applications:

The passage introduces the perspective of Chester Drum, a doctor and co-founder of AOK-pass, who believes that proof of health status will become a common requirement for various activities. It mentions the facilitation of processes through digital health passports.

Move 6 Addressing Security Measures:

This move discusses the advantage of using blockchain technology in preventing forgeries through tamper-proof test results and vaccination certificates. It emphasizes the importance of security and trust in the system.

Move 7 Acknowledging Privacy Concerns:

The passage raises concerns about data privacy and leaks, citing the need for appropriate measures to protect users' data and maintain the confidentiality of test results. It mentions instances of social exclusion due to the disclosure of positive test results.

Move 8 Highlighting AOK-pass's Approach:

This move assures readers that AOK-pass prioritizes privacy by encrypting personal health information and claims not to monetize user data. It mentions the company's contracts with airports and airlines for testing and development.

Move 9 Addressing Challenges:

The passage acknowledges the challenges of achieving seamless integration of digital health passports throughout the travel process. It mentions the coordination required among different entities and the complexities arising from varying regulations and procedures.

Move 10 Providing Personal Experience:

The passage concludes by sharing Giulio Castagnara's experience during his layover in New York, where the proof of health status was not requested. It highlights his perception of digital health passports as a game changer and his appreciation for the added assurance of safety while traveling.

Communicative Purpose:

The purpose of the passage is to inform readers about the concept of digital health passports, particularly AOK-pass, and their implementation in the travel industry. It aims to highlight the benefits, challenges, and potential impact of digital health passports on travel safety and the return to pre-pandemic activities. The passage also addresses concerns related to data privacy and showcases the perspectives of experts and individuals involved in the development and implementation of digital health passports.

Culture Tips

1. Familiarize yourself with local regulations: Before traveling to a specific destination, make sure to research and understand the local regulations regarding digital health passports or any other health verification requirements. Stay updated on the latest information to ensure a smooth travel experience.

2. Be prepared: Download the necessary digital health passport app recommended or required by the airport or airline you'll be traveling with. Familiarize yourself with the app's features and ensure you have all the required information, such as test results or vaccination records, readily available.

3. Follow instructions: Listen carefully to instructions provided by airport personnel or airline staff regarding the use of digital health passports. They will guide you through the process and ensure compliance with health and safety protocols.

4. Respect privacy and security measures: While using digital health passports, follow the privacy and security guidelines suggested by the app or provided by the relevant authorities. Be cautious with sharing personal information and keep your phone and app secure to protect your data.

5. Patience and understanding: Understand that the implementation of digital health passports is still relatively new, and there may be some variations in procedures across different airports or airlines. Stay patient and cooperative, knowing that these measures are in place to prioritize everyone's safety.

6. Stay informed: Keep yourself updated on the latest developments and changes in digital health passport requirements, both in your home country and in the destinations you plan to travel to. This will help you stay prepared and minimize any potential surprises or inconveniences.

7. Respect local customs and etiquette: Remember that digital health passports are just one aspect of travel. Respect the local customs, cultural norms, and etiquette of the places you visit. Stay informed about any specific health and safety guidelines in each destination and adhere to them accordingly.

8. Be courteous to airport and airline staff: Airport personnel and airline staff are working hard to ensure a safe and smooth travel experience for everyone. Show appreciation for

their efforts by being polite, patient, and cooperative throughout the process.
9. Remember, cultural norms and practices may vary, so it is essential to adapt and respect the specific guidelines of each country or region you visit.

Reading Exercises

I. Vocabulary Matching

Directions: *Fill in the blank with the corresponding letter of the Chinese equivalent of the following English words.*

1. board ( )   8. estimate ( )   15. prevent ( )   22. launch ( )
2. download ( )   9. departure ( )   16. essential ( )   23. coordinate ( )
3. digital ( )   10. available ( )   17. privacy ( )   24. immigration ( )
4. verify ( )   11. status ( )   18. exclude ( )   25. staff ( )
5. reliable ( )   12. procedure ( )   19. recommend ( )
6. print ( )   13. arrest ( )   20. maintain ( )
7. interpret ( )   14. individual ( )   21. schedule ( )

A. 移民，移居   H. 估计   O. 印刷   V. 程序，手续
B. 发行   I. 离开，起程   P. 逮捕，拘留   W. 登机（或上船）
C. 使协调   J. 可获得的   Q. 数码的，数字式的 X. 个别的
D. 阻止，阻碍   K. 推荐，举荐   R. 证实，证明   Y. 诠释，说明
E. 极其重要的   L. 维持，保持   S. 可信赖的
F. 隐私，私密   M. 地位，身份   T. 安排
G. 不包括   N. 全体职工   U. 下载

II. Answer the Following Questions

1. What is AOK-pass, and how does it work?
2. Why are airports and airlines experimenting with digital health passports like AOK-pass?
3. How does AOK-pass prevent counterfeiting of COVID-19 test results?
4. What concerns are associated with the use of digital health passports?
5. How does AOK-pass plan to launch and compete in the market?

III. Multiple Choice

1. What is AOK-pass? ( )

    A. A digital health passport.

    B. An airline company.

C. A COVID-19 testing site.

D. A vaccination certification.

2. What is the purpose of a digital health passport like AOK-pass?（　）

   A. To simplify the process of verifying COVID-19 test results or vaccination status.

   B. To replace traditional paper immunization cards.

   C. To monetize user data for profit.

   D. To eliminate the need for COVID-19 testing before travel.

3. How does the AOK-pass app generate proof of a traveler's health status?（　）

   A. By scanning the traveler's passport.

   B. By analyzing the traveler's symptoms.

   C. By generating a unique QR code.

   D. By conducting an on-site COVID-19 test.

4. What advantage does using blockchain technology provide for digital health passports?（　）

   A. It speeds up the verification process at airports.

   B. It eliminates the need for COVID-19 testing.

   C. It prevents forgeries and enhances security.

   D. It monetizes user data for profit.

5. What challenges exist in implementing digital health passports throughout the entire travel process?（　）

   A. Varying regulations and procedures among different entities.

   B. Lack of available COVID-19 testing sites.

   C. Insufficient smartphone compatibility.

   D. Limited funding for app development.

Translation

Directions: *Translate the following sentences into Chinese.*

1. Travel industry experts say this type of digital health certificate could be a game changer, not just for tourism, but live events and schools.

   _____

   _____

2. In the setting of COVID-19, health information is beginning to be used in a transitional

sense, much more than before.

3. Let me show you part of my medical record to get on an airplane.

4. Policymakers should require digital health password providers to adopt appropriate technical and organizational measures to address potential data privacy invasive situations practically.

5. What's hard is the coordinated process, starting with a laboratory in a different country, creating a foreign medical record, boarding an airline, another highly regulated industry, and then crossing an immigration entry, which itself is regulated by the rules and regulations of a country different than where you got your health test.

Speaking

Traveler (Giulio Castagnara): Excuse me, I have the AOK-pass app on my phone. Can you scan my QR code for verification?

Airport personnel: Of course, sir. Please hold your phone steady, and I'll scan the QR code right away.

(Airport personnel scans the QR code on Giulio's phone.)

Airport personnel: Thank you, Mr. Castagnara. Your health status has been verified, and you are cleared to proceed with check-in and boarding.

Giulio Castagnara: Great, thank you! It's reassuring to know that the digital health passport is being used to ensure everyone's safety.

Airport personnel: Absolutely. It's part of our efforts to prioritize the well-being of our passengers and staff. If you have any other questions or need further assistance, please don't hesitate to ask.

Giulio Castagnara: Actually, I have a layover in New York before my final destination in

Dallas. Will I need to go through the health verification process again there?

Airport personnel: The implementation of the digital health passport may vary between airports, but it's always a good idea to have your AOK-pass app ready. While I can't speak specifically for the airport in New York, it's possible they might require verification. I recommend keeping your QR code easily accessible to streamline the process.

Giulio Castagnara: Understood. I'll make sure to have it readily available. Thank you for the information.

Airport personnel: You're welcome, Mr. Castagnara. Have a pleasant flight, and if you have any other questions during your journey, don't hesitate to reach out to the airline or airport staff at your next destination.

Giulio Castagnara: I appreciate your assistance. Looking forward to a smooth and safe travel experience. Thank you again.

Airport personnel: It's our pleasure, Mr. Castagnara. Safe travels and enjoy your trip.

Writing

Directions: *For this part, you are allowed 30 minutes to write an essay about the importance of developing a healthy lifestyle among college students. You should write at least 120 words but no more than 180 words.*

_____
_____
_____
_____
_____
_____
_____

Expansion Exercise for CET-BAND 4—Reading

Directions: *The passage is followed by some questions or unfinished statements. For each of them there are four choices marked A, B, C and D. You should decide on the best choice.*

Three children in every classroom have a diagnosable mental health condition. Half of these are behavioural disorders, while one-third are emotional disorders such as stress, anxiety and depression, which often become outwardly apparent through self-harm. There was an astonishing 52 percent jump in hospital admissions for children and young people who had harmed themselves between 2009 and 2015.

Schools and teachers have consistently reported the scale of the problem since 2009. Last year over half of teachers reported that more of their pupils experience mental health problems than in the past. But teachers also consistently report how ill-equipped they feel to meet pupils' mental health needs, and often cite a lack of training, expertise and support from the National Health Service（英国国家医疗服务体系）.

Part of the reason for the increased pressure on schools is that there are now fewer early intervention（干预）and low-level mental health services based in the community. Cuts to local authority budgets since 2010 have resulted in a significant decline of these services, despite strong evidence of their effectiveness in preventing crises further down the line.

The only way to break the pressures on both mental health services and schools is to reinvest in early intervention services inside schools.

There are strong arguments for why schools are best placed to provide mental health services. Schools see young people more than any other service, which gives them a unique ability to get to hard-to-reach children and young people and build meaningful relationships with them over time. Recent studies have shown that children and young people largely prefer to see a counsellor in school rather than in an outside environment. Young people have reported that for low-level conditions such as stress and anxiety, a clinical setting can sometimes be daunting（令人却步的）.

There are already examples of innovative schools which combine mental health and wellbeing provision with a strong academic curriculum. This will, though, require a huge **cultural shift**. Politicians, policymakers, commissioners and school leaders must be brave enough to make the leap towards reimagining schools as providers of health as well as education services.

1. What are teachers complaining about?（   ）

   A. There are too many students requiring special attention.

   B. They are under too much stress counselling needy students.

   C. Schools are inadequately equipped to implement any intervention.

   D. They lack the necessary resources to address pupils' mental problems.

2. What do we learn from the passage about community health services in Britain? (   )

   A. They have deteriorated due to budget cuts.

   B. They facilitate local residents' everyday lives.

   C. They prove ineffective in helping mental patients.

   D. They cover preventative care for the local residents.

3. Where does the author suggest mental health services be placed? (   )

   A. At home.

   B. At school.

   C. In hospitals.

   D. In communities.

4. What do we learn from the recent studies? (   )

   A. Students prefer to rely on peers to relieve stress and anxiety.

   B. Young people are keen on building meaningful relationships.

   C. Students are more comfortable seeking counselling in school.

   D. Young people benefit from various kinds of outdoor activities.

5. What does the author mean by a cultural shift (Line 2, Para. 6) ? (   )

   A. Simplification of schools' academic curriculums.

   B. Parents' involvement in schools' policy-making.

   C. A change in teachers' attitudes to mental health.

   D. A change in the conception of what schools are.

Unit 2 Aviation Science

Preview

Airplane flight involves the interaction of various forces of physics. The four forces acting on an airplane include weight, lift, drag, and thrust. Lift, generated primarily by the wings, counters the force of gravity, while thrust from the engines propels the plane forward. Commercial jets can reach speeds over 600 miles per hour and fly at high altitudes, while supersonic planes travel faster than the speed of sound. However, supersonic flight creates disruptive sonic booms, which are loud shockwaves caused by the sudden change in air pressure. Researchers have worked on reducing the intensity of sonic booms to make supersonic flight more acceptable. Although the sonic boom problem is yet to be fully resolved, scientists are making progress towards developing quieter supersonic aircraft. The ultimate goal is to enhance society's benefits through scientific advancements in aviation.

Listening

Directions: *Fill in the blanks with the words that you hear. Find them in the table below and write down the corresponding letters.*

There are many invisible __1__ of physics that played during each flight. So let's start with the __2__. How does physics play into airplane flight? The four forces acting on an airplane in flight include its weight which acts downward to pull it back down to the earth. That force has to be overcome by __3__ lift which lifts the aircraft and enables it to go to altitude. Lift is the __4__ force produced by an aircraft, __5__ the wing and it's influenced by the shape of the wing. Another important force __6__ on the airplane is what we call aerodynamic drag that's caused by the __7__ of the air. As the aircraft flies __8__ the air, it doesn't __9__ it, just resists the forward motion of the airplane, and that force of course has to be balanced by thrust which is produced by the __10__.

| A. primarily | B. forces | C. acting | D. resistance | E. vertical |
| F. through | G. aerodynamic | H. engines | I. basics | J. pull |

Reading

The Science of Airplane Flight

Section A

Pilot: **Fasten** your **seatbelts**. You're about to take to the skies.

ATC: Flight Zero Zero One, you're **clear** for takeoff.

Pilot: **Roger** that.

Airplane flight is more popular than ever. Millions of **passengers** fly each year. Pilots make fly look easy, but it's not. There are many **invisible forces** of **physics** that played during each flight. How does it all work? To find out, let's talk to some scientists at NASA with the National **Aeronautics** and Space **Administration**.

Unit 2 Aviation Science

Section B

"Hi, my name is Joe Chambers. I'm a **retired aeronautical engineer**, retired from the NASA Langley Research Center, after a 36-year-**career** in aeronautics." Aeronautics is the science of airplane flight. "As I find that children at your age ask questions that in many **cases** are more difficult to answer than adults." So let's start with the basics. How does physics play into airplane flight?

"The four forces acting on an airplane in flight include its weight which acts downward to pull it back down to the earth. That force has to be overcome by **aerodynamic lift** which lifts the aircraft and enables it to go to altitude. Lift is the **vertical** force produced by an aircraft, **primarily** the wing, and it is influenced by the shape of the wing. Another important force acting on the airplane is what we call **aerodynamic drag** that is caused by the **resistance** of the air as the aircraft flies through the air. It doesn't **pull**; it just resists the forward **motion** of the airplane. And that force, of course, has to be balanced by **thrust** which is produced by the **engines**." Thrust makes the engines of a plane go fast.

Section C

Commercial jets can fly at over 600 miles an hour. They fly at pretty high altitudes too, at over 30,000 feet above the ground. Some planes go even faster, they're called **supersonic** planes because they fly faster than the speed of sound which is about 760 miles per hour at sea level.

Section D

"We call the fastest planes supersonic airplanes because they travel faster than the speed of sound and they get people to their destination a lot faster." "My name is Christine Darden and I work at NASA Langley Research Center for nearly 40 years. The problem that I have worked on in my career is that airplanes that travel faster than the speed of sound **create** a **sonic boom** that people on the ground can hear. Sonic booms are really loud. It sounds like a boom or a loud **crack**. And my work was to try to see if we could reduce that sonic boom, so they would not be this **disturbing byproduct** of supersonic flight."

Section E

We usually don't see supersonic planes in the air today because of the sonic boom noise. A sonic boom is a **shockwave** that works like this—"as you **blow up** a balloon, you're blowing into it the pressure, it's getting higher and higher into the balloon. And that's of course why it gets bigger because you're getting more **molecules** inside the balloon. If I took a needle or a pen and **pop** that balloon, there is an **immediate** shockwave set up in all **directions**. It's like a **sphere** going out from that balloon in all directions with this high pressure air, and so, it is going toward you. As this **boundary** of  this sphere goes across your ear, you're getting an **instantaneous** change in pressure, from the normal pressure to the high pressure that was in the balloon. But that change in pressure **instantaneously** is what you hear.

Section F

"And that's the same thing that happens with an airplane when it is going faster than the speed of sound. It's almost like an ice-cream **cone** on the nose of the airplane, and all of the molecules that have been **disturbed** by that airplane are **pushed out of** the way, and the higher pressure air is all within that cone, and this cone goes all the way to the ground. So you hear that instantaneous change between the normal air and the air that has been inside that cone, and that's what you hear when you hear a sonic boom."

Section G

"One of the **approaches** we took to trying to change the sonic boom was to see if **shaping** the airplane differently would **impact** the sonic boom, and so sometimes you find the airplane might be longer. One of the ways that we **test out** our ideas about how to reduce the sonic boom is to build models of the designs that we think might work. And we would put that model in a **wind tunnel**, and **measure** the pressures coming off that model in the wind tunnel. The air is actually **blown** past the model itself, which gives the pretty much the same effects. And we measure data points that we want inside the wind tunnel. That's a lot cheaper than building a big airplane and flying it through the air, which is **prohibitively** expensive."

Section H

Wind tunnels have been popular ways to test airplanes since the 1930s. Many tests are done at NASA's Langley Research Center in Hampton Virginia. "One of the most famous wind tunnels was the Langley **full-scale** wind tunnel which was **enclosed** in the building

as long as a football field. And it had a test **section** where full-scale aircraft were tested. The test section measured 60 feet across and 30 feet high, which is **immense**, ran at an airspeed of a hundred miles an hour and **operated** for over 80 years."

Section I

Other wind tunnels are smaller and the air travels faster. Scientists use the smaller wind tunnels to test new supersonic plane designs. They haven't solved the sonic boom problem yet, but they're getting close. "We're hoping for a lower **rumble** (sonic) beam that would not upset people. That's **ultimately** the goal that you want to benefit society with the scientific **principles** and advances, you want to show how that benefits, and how we live."

So, **kick back**, relax and enjoy the ride.

New Words and Expressions

▲**fasten** *v.* /ˈfɑːsn/   to close or join together the two parts of sth.
（使两部分）系牢，扎牢，结牢，扣紧

seatbelt *n.* /ˈsiːtbelt/   a safety belt used in a car or plane to hold you in your seat in case of an accident
安全带

clear *adj.* /klɪr/   free from things that are blocking the way or covering the surface of sth.
畅通无阻的，无障碍的，（表面）收拾干净的

roger *v.* /ˈrɒdʒər/   to show that they have understood a message
（用于无线电通讯）信息收到，明白

passenger *n.* /ˈpæsɪndʒər/   a traveler riding in a vehicle
乘客，旅客

invisible *adj.* /ɪnˈvɪzəbl/   not prominent or readily noticeable
看不见的，隐形的，无形的

force *n.* /fɔːrs/   an effect that causes things to move in a particular way
[物]力

physics *n.* /ˈfɪzɪks/	the scientific study of forces such as heat, light, sound, etc., of relationships between them, and how they affect objects 物理学
aeronautics *n.* /ˌerəˈnɔːtɪks/	the science or practice of building and flying aircraft 航空学，飞行学，飞行术
▲**administration** *n.* /ədˌmɪnɪˈstreɪʃn/	the process of organizing and supervising sth. 管理，监管
NASA *n.* /ˈnæsə/	National Aeronautics and Space Administration, a US government organization that does research into space and organizes space travel （美国）国家航空和航天局
retired *adj.* /rɪˈtaɪərd/	no longer active in your work or profession 已退休的，已退职的
aeronautical *adj.* /ˌerəˈnɔːtɪkl/	of or pertaining to aeronautics 航空（学）的，飞机设计制造的
engineer *n.* /ˌendʒɪˈnɪr/	a person whose job is to design and build engines, machines, roads, bridges, etc.; or control and repair machines and equipments 工程师，设计师，机修工，技师，（船上的）轮机手，（飞机上的）机械师
career *n.* /kəˈrɪr/	the general progression of your working or professional life 生涯，职业，经历，事业
case *n.* /keɪs/	a special set of circumstances 具体情况，事例
aerodynamic *adj.* /ˌerəʊdaɪˈnæmɪk/	of or relating to the qualities of an object that affect the way it moves through the air （与）空气动力学（有关）的 of or having a shape which reduces the drag from air moving past

	（减少空气阻力的）流线型的，具有流线型构造的
▲vertical *adj.* /ˈvɜːrtɪkl/	straight up or down 竖的，垂直的，直立的
primarily *adv.* /praɪˈmerəli/	for the most part, mainly 主要地，根本地
resistance *n.* /rɪˈzɪstəns/	a mechanical force that tends to retard or oppose motion 抵抗力，阻力
pull *v.* /pʊl/	to hold sth. firmly and use force to move it towards yourself 拉，拽，扯，拖
motion *n.* /ˈmoʊʃn/	the act or process of moving or the way sth. moves 运动，移动，动
thrust *n.* /θrʌst/	the force that is produced by an engine to push a plane, rocket, etc. forward （发动机推动飞机、火箭等的）推力，驱动力
engine *n.* /ˈendʒɪn/	motor that converts thermal energy to mechanical work 发动机，引擎
supersonic *adj.* /ˌsuːpərˈsɒnɪk/	(of speed) greater than the speed of sound in a given medium (especially air) 超音速的，超声速的
create *vt.* /kriˈeɪt/	to make sth. happen or exist 创造，创作，创建
sonic *adj.* /ˈsɒnɪk/	(technical) connected with sound or the speed of sound （术语）声音的，声速的
boom *n.* /ˈbuːm/	a loud, deep, resonant sound 低沉的声音，隆隆声
▲crack *n.* /kræk/	a sudden sharp loud noise （突然的）爆裂声，噼啪声，轰响
disturbing *adj.* /dɪˈstɜːrbɪŋ/	causing distress, worry or anxiety 引起烦恼的，令人不安的，引起恐慌的

byproduct *n.* /ˈbaɪprɒdəkt/	a secondary and sometimes unexpected consequence 副产品，意外收获，附带的结果
shockwave *n.* /ˈʃɑːkweɪv/	a movement of very high air pressure that is caused by an explosion, earthquake, etc. （爆炸、地震等引起的）冲击波
molecule *n.* /ˈmɒlɪkjuːl/	the simplest structural unit of an element or compound 分子
pop *v.* /pɒp/	cause to burst with a loud, explosive sound; to put sth. somewhere quickly, suddenly or for a short time （使）爆裂，发出爆裂声，（迅速或突然）放置
immediate *adj.* /ɪˈmiːdiət/	happening or done with little or no delay 立即的，立刻的
direction *n.* /dəˈrekʃn/	the general position a person or thing moves or points towards 方向，方位
sphere *n.* /sfɪr/	a three-dimensional closed surface such that every point on the surface is equally distant from the center 球，球体，球形，球状物
boundary *n.* /ˈbaʊndri/	a real or imagined line that marks the limits or edges of sth. and separates it from other things or places; a dividing line 边界，界限，分界线
instantaneous *adj.* /ˌɪnstənˈteɪniəs/	occurring with no delay 立即的，立刻的，瞬间的
instantaneously *adv.* /ˌɪnstənˈteɪniəsli/	without any delay 瞬间，即刻，突如其来地
cone *n.* /kəʊn/	a solid or hollow object with a round flat base and sides that slope up to a point; a kind of ice-cream （实心或空心的）圆锥体，锥形蛋卷冰激凌

disturb *v.* /dɪˈstɜːrb/	interfere with the normal arrangement or functioning of 打扰，干扰，妨碍，搅乱
▲**approach** *n.* /əˈprəʊtʃ/	ideas or actions intended to deal with a problem or situation 方法，建议
shape *v.* /ʃeɪp/	to make sth. into a particular look 使成为……形状（或样子），塑造
impact *v.* /ɪmˈpækt/	to have an effect on sth. （对某事物）有影响，有作用
measure *v.* /ˈmeʒər/	to find the size, quantity, etc. of sth. in standard units 测量，度量
blow *v.* /bləʊ/	cause air to go in, on, or through 吹，刮，（被）刮动，吹动
prohibitively *adv.* /prəʊˈhɪbətɪvli/	to an extreme degree 过高地，过分地，禁止地
full-scale *adj.* /ˌfʊl ˈskeɪl/	as complete and thorough as possible; the same size as sth. that is being copied 全面的，完全的，原尺寸的，和实物同样大小的
enclose *v.* /ɪnˈkləʊz/	to surround sth. （用墙、篱笆等）把……围起来，围住
section *n.* /ˈsekʃn/	any of the parts into which sth. is divided 部分，部门
immense *adj.* /ɪˈmens/	extremely large or great, especially in scale or degree 极大的，巨大的
operate *v.* /ˈɒpəreɪt/	perform as expected when applied 工作，操作，控制，使运行
rumble *n.* /ˈrʌmbl/	a long deep sound or series of sounds 持续而低沉的声音，隆隆声
▲**ultimately** *adv.* /ˈʌltɪmətli/	as the end result of a succession or process 最终，最后，终归，最基本地，根本上
principle *n.* /ˈprɪnsəpl/	a law, a rule or a theory that sth. is based on

	法则，原则，原理
aerodynamic lift	［航］气动升力，空气动力升力
aerodynamic drag	［航］气动阻力，空气动力阻力
wind tunnel	（试验飞机等用的）风洞，风道
blow up	爆发，爆炸，放大，使充气
push out of	把……推出
test out	彻底检验，考验
as long as	只要，和……一样长
kick back	放松

Genre Analysis

Genre:

The passage can be classified as an informative article or educational piece discussing the science of airplane flight, including the forces involved, the concept of sonic booms, and the use of wind tunnels for testing.

Communicative Moves:

Move 1 Setting the Scene:

The passage begins with a dialogue between a pilot and air traffic control, creating a sense of anticipation and introducing the topic of airplane flight.

Move 2 Introducing the Expert:

The passage introduces Joe Chambers, a retired aeronautical engineer, who explains the role of physics in airplane flight. He discusses the four forces acting on an airplane and how they are balanced.

Move 3 Highlighting Speed and Altitude:

This move provides information about the speed and altitude at which commercial jets can fly, including supersonic planes that travel faster than the speed of sound.

Move 4 Addressing Sonic Booms:

Christine Darden, a NASA researcher, discusses the issue of sonic booms created by supersonic planes and her work in reducing their impact.

Move 5 Explaining Sonic Booms:

This move provides a simple analogy of a balloon popping and the resulting shockwave to explain how sonic booms are formed.

Move 6 Describing the Cone Effect:

The passage uses an analogy of an ice cream cone to describe the shape of the high-pressure air around a supersonic plane. It explains how the change in pressure creates the sound of a sonic boom.

Move 7 Discussing Approaches to Reduce Sonic Booms:

This move explains the efforts made to change the shape of airplanes to minimize sonic booms. It mentions the use of wind tunnels and model testing as cost-effective alternatives to full-scale aircraft testing.

Move 8 Highlighting Wind Tunnel Testing:

The passage discusses the history and significance of wind tunnels, particularly the Langley full-scale wind tunnel, in testing aircraft designs. It emphasizes the size and longevity of wind tunnels.

Move 9 Expressing Progress and Goals:

This move acknowledges that the sonic boom problem has not been completely solved but mentions the ongoing efforts to develop quieter supersonic planes. It emphasizes the goal of benefiting society through scientific advancements.

Move 10 Concluding on a Positive Note:

The passage concludes by encouraging readers to enjoy the flight and implying that advancements in the science of airplane flight will continue to enhance the travel experience.

Communicative Purpose:

The purpose of the passage is to provide an educational overview of the science behind airplane flight. It aims to explain the forces involved in flight, introduce the concept of sonic booms, and discuss the efforts made by scientists and engineers to reduce their impact. The passage also highlights the use of wind tunnels for testing aircraft designs and emphasizes the ongoing progress in the field. The overall goal is to inform and engage readers, sparking their curiosity about the science and engineering behind aviation.

Culture Tips

1. Appreciate scientific advancements: The passage highlights the ongoing progress in the field of aviation. Culturally, it's important to appreciate and celebrate scientific advancements that enhance our understanding of the world around us. Encourage a curiosity for science and engineering and acknowledge the hard work of researchers and engineers who contribute to technological advancements.

2. Foster a sense of wonder: The passage emphasizes the awe-inspiring nature of airplane flight. Culturally, it's beneficial to nurture a sense of wonder and curiosity about the world. Encourage exploration and encourage individuals to ask questions and seek

knowledge about the science behind everyday phenomena.

3. Support STEM education: The passage showcases the importance of STEM (Science, Technology, Engineering, and Mathematics) fields in aviation. Culturally, it's essential to support and promote STEM education, especially among younger generations. Encourage students to pursue studies and careers in science and engineering, as they contribute to society's progress and technological development.

4. Environmental awareness: The passage briefly touches on the issue of sonic booms created by supersonic planes. Culturally, it's important to prioritize environmental considerations in technological advancements. Encourage sustainable practices and support research and development aimed at reducing the environmental impact of transportation technologies.

5. Embrace multidisciplinary collaboration: The passage mentions the collaboration between aeronautical engineers, NASA researchers, and air traffic control. Culturally, it's crucial to embrace multidisciplinary collaboration to solve complex problems. Encourage collaboration between different fields of expertise, as diverse perspectives often lead to innovative solutions.

6. Celebrate cultural diversity: The passage doesn't explicitly mention cultural diversity, but aviation and scientific advancements are the result of collaborative efforts from people of various backgrounds and cultures. Culturally, it's important to celebrate and embrace diversity, as it brings unique perspectives and ideas to the table. Encourage inclusivity and respect for different cultures within the scientific and engineering communities.

Reading Exercises

I. Vocabulary Matching

Directions: *Fill in the blank with the corresponding letter of the Chinese equivalent of the following English words.*

1. force   (   )
2. supersonic   (   )
3. prohibitively   (   )
4. direction   (   )
5. aerodynamic   (   )
6. vertical   (   )
7. boundary   (   )
8. sonic boom   (   )
9. engineer   (   )
10. aeronautics   (   )
11. resistance   (   )
12. full-scale   (   )
13. instantaneous   (   )
14. wind tunnel   (   )
15. sphere   (   )
16. career   (   )
17. retired   (   )
18. disturb   (   )
19. cone   (   )
20. physics   (   )
21. shockwave   (   )
22. engine   (   )
23. molecule   (   )
24. passenger   (   )

A. 打扰，干扰　　G. 乘客，旅客　　M. 球，球体　　S. 圆锥体，蛋筒
B. 禁止地　　　　H. 超音速的　　　N. 立即的，立刻的　T. 边界，界限
C. 工程师，设计师　I. 竖的，垂直的　O. 全面的，完全的　U. 生涯，职业
D. 方向，方位　　J. 航空学　　　　P. 风洞，风道　　V. 物理学
E. 空气动力学的　K. 力，力量　　　Q. 抵抗力，阻力　W. 音爆，声爆
F. 冲击波，震惊　L. 分子　　　　　R. 发动机　　　　X. 已退休的

II. Answer the Following Questions

1. What are the four forces acting on an airplane during flight?
2. What is the primary purpose of aerodynamic lift?
3. What is the speed of sound at sea level?
4. What causes a sonic boom?
5. How do scientists test new supersonic plane designs without building full-scale aircraft?

III. Multiple Choice

1. What is the primary force that lifts an airplane during flight?（　）

    A. Weight.

    B. Drag.

    C. Thrust.

    D. Lift.

2. At what altitude do commercial jets typically fly?（　）

    A. Below 10,000 feet.

    B. Around 20,000 feet.

    C. Over 30,000 feet.

    D. Above 40,000 feet.

3. What is the main challenge associated with supersonic flight?（　）

    A. Maintaining altitude.

    B. Reducing noise from sonic booms.

    C. Achieving higher speeds.

    D. Handling increased air resistance.

4. How do scientists at NASA test new supersonic plane designs?（　）

    A. By conducting test flights with full-scale aircraft.

    B. By using computer simulations.

    C. By building wind tunnels.

    D. By analyzing data from previous supersonic flights.

5. What is the goal of reducing the sonic boom noise associated with supersonic flight?（　）

A. To improve fuel efficiency.

B. To increase passenger comfort.

C. To enhance maneuverability.

D. To decrease air pollution.

Translation

Directions: *Translate the following sentences into Chinese.*

1. There are many invisible forces of physics that played during each flight.

2. I'm a retired aeronautical engineer, retired from the NASA Langley Research Center, after a 36-year-career in aeronautics.

3. The aerodynamic drag doesn't pull; it just resists the forward motion of the airplane.

4. My work was to try to see if we could reduce that sonic boom, so they would not be this disturbing byproduct of supersonic flight.

5. It's like a sphere going out from that balloon in all directions with this high pressure air.

6. It's almost like an ice-cream cone on the nose of the airplane, and all of the molecules that have been disturbed by that airplane are pushed out of the way, and the higher pressure air is all within that cone, and this cone goes all the way to the ground.

7. (In wind tunnels) The air is actually blown past the model itself, which gives the pretty much the same effects. And we measure data points that we want inside the wind tunnel.

Unit 2 Aviation Science

8. One of the most famous wind tunnels was the Langley full-scale wind tunnel which was enclosed in the building as long as a football field. And it had a test section where full-scale aircraft were tested.

9. We're hoping for a lower rumble (sonic) beam that would not upset people.

10. That's ultimately the goal that you want to benefit society with the scientific principles and advances, you want to show how that benefits, and how we live.

Speaking

Student: Hi, my name is Sarah. I've always been fascinated by airplanes and how they fly. Can you explain to me how physics plays a role in airplane flight?

Expert: Hi Sarah, I'm Joe Chambers, a retired aeronautical engineer with NASA. I'd be happy to explain it to you. Physics is indeed at the heart of airplane flight. Let's start with the four forces that act on an airplane during flight.

Student: That sounds interesting. What are these four forces?

Expert: The four forces are weight, lift, drag, and thrust. Weight is the force pulling the airplane down towards the Earth. Lift, on the other hand, is the upward force generated by the wings that counteracts the weight and allows the airplane to stay airborne. Drag is the resistance caused by the air as the airplane moves through it. Finally, thrust is the force produced by the engines that propels the airplane forward.

Student: So, to keep the airplane flying, the lift force must be greater than the weight force, right?

Expert: Exactly! For the airplane to achieve and maintain flight, the lift force needs to be equal to or greater than the weight force. This balance allows the airplane to stay at a particular altitude.

Student: That makes sense. But what about drag? How does that affect the airplane?

Expert: Drag is indeed a significant factor. It resists the forward motion of the airplane. To overcome drag, the airplane needs thrust generated by the engines. Thrust propels the airplane forward, overcoming the resistance caused by drag. It's like pushing through the air to maintain the desired speed.

Student: I see. That explains how airplanes maintain their speed and altitude. But what about those supersonic planes that travel faster than the speed of sound? How do they create those sonic booms?

Expert: Ah, great question. When an airplane exceeds the speed of sound, it creates a shockwave called a sonic boom. It happens because the airplane is pushing through the air faster than the speed at which sound waves travel. This creates a sudden change in pressure, which we hear as a loud boom.

Student: That's fascinating! Is there any way to reduce or eliminate the sonic booms?

Expert: Yes, indeed. Scientists, including myself, have been working on finding ways to minimize sonic booms. By shaping the airplane differently, we can influence the formation and strength of the shockwave. We use wind tunnels to test different designs and measure the pressures generated by the models. This helps us understand how to reduce the disruptive effects of sonic booms, making supersonic flight quieter.

Student: Wind tunnels sound amazing! So, by testing in wind tunnels, you can gather data without the need for expensive full-scale aircraft testing?

Expert: Precisely! Wind tunnels provide an efficient and cost-effective way to simulate flight conditions and evaluate aircraft designs. It allows us to measure important data points without the need for extensive real-world testing, which can be quite expensive.

Student: That's incredible. I never realized there was so much science and engineering involved in airplane flight. Thank you for sharing all this information, Mr. Chambers!

Expert: You're welcome, Sarah. I'm glad I could help you understand the fascinating world of airplane flight. It's a field that continues to evolve, and who knows what exciting developments lie ahead. If you have any more questions, feel free to ask.

Writing

Directions: *For this part you are allowed 30 minutes to write an essay on whether technology will make people lazy. You should write at least 120 words but no more than 180 words.*

Expansion Exercise for CET-BAND 4—Reading

Directions: *The passage is followed by some questions or unfinished statements. For each of them there are four choices marked A, B, C and D. You should decide on the best choice.*

In the age of the Internet, there's no such thing as a private debate. But is that bad for science? Some scientists have had concerns. When debates in any sector move beyond the halls of universities and government agencies, there's potential for information to be used incorrectly, leading to public confusion; yet, open debate can also promote communication between the scientific community and the public. Recent open debates on scientific research, health, and policy have aroused greater public attention and encouraged more diverse voices. If this trend spurs scientists to agree more quickly about the best solutions to our problems—and at the same time helps the public observe the process of scientific discourse more clearly—then this is good for everyone, including scientists.

A recent debate published in The New York Times discussed the question of how quickly medicine should be developed and produced. Issues such as safety of the product and perception of the public were examined and considered. But some experts worried that such public speculation might lead people to believe that disagreement about the details meant a lack of adequate scientific consensus over the safety and efficiency of modern-day medicine.

The anxiety seems misplaced. Gone are the days of going to a conference and debating scientific issues, and that's good because those gatherings were not diverse enough and excluded many important voices. These days, the public can access debates about science regardless of where they take place.

For many scientists, public debate is a new frontier and it may feel like a place with few restraints or rules. But rather than avoiding such conversations, let the debates be transparent and vigorous, wherever they are held. If the public is to understand that science is an honorably self-correcting process, the idea that science is a fixed set of facts in a textbook needs to be dismissed. With the validity of science coming under attack, there's a need for scientific debates to be perceived as open and true to life. Let everyone see the noisy, messy deliberations that advance science and lead to decisions that benefit us all.

1. What does the author think open debate can do? （　）

    A. Help the public to better understand science.

    B. Clear up confusion in the scientific community.

    C. Settle disputes between universities and government agencies.

    D. Prevent information from being used incorrectly by the public.

2. Why did a recent debate published in The New York Times arouse concerns among experts? （　）

    A. It might hinder the progress in medical research.

    B. It might breed public distrust in modem medicine.

    C. It might add to the difficulty of getting research funds.

    D. It might prevent medical scientists reaching consensus.

3. Why does the author say some experts' anxiety seems misplaced? （　）

    A. Debating scientific issues at a conference is now old-fashioned.

    B. Diverse topics can be debated by both scientists and the public.

    C. Debates about science are accessible to the public anyway.

    D. Scientists can voice their opinions whatever way they like.

4. What does the author suggest scientists do about public debate? （　）

    A. Have more discussions about it.

    B. Embrace it with open arms.

    C. Formulate new rules for it.

    D. Restrain it to a rational degree.

5. What does the author say about science in the last paragraph? （　）

    A. It is transmitted through textbooks.

    B. It is what proves valid and true to life.

    C. It is a dynamic and self-improving process.

    D. It is a collection of facts and established rules.

Unit 3 Aviation Humanity

Preview

　　The passage highlights the importance of human factors training in various organizations by drawing parallels to the components of an airplane. It emphasizes the need for well-trained employees (wings) and motivated individuals (engines) to propel the organization forward. Effective communication is compared to fuel, essential for smooth operations. Control, coordination, and anticipation of obstacles (rudder and elevator) are crucial for a successful journey. The radar represents the ability to identify and avoid external threats. Training for complex situations (landing gear) and emergency preparedness ensure safety and efficiency. The passage concludes by emphasizing the ongoing need for effective teamwork and individual contributions in overcoming future obstacles.

Listening

Directions: *Fill in the blanks with the words that you hear. Find them in the table below and write down the corresponding letters.*

It is unfortunately also full of __1__ and its paths are often complex. The issue of safety is always with us and we do our best to protect ourselves from known risks. Most of the time the sun shines and people go about their lives __2__. However, clouds can form quickly. We have developed tools to overcome obstacles quickly, __3__ and safely. Most of you have probably heard of it. It's called an __4__.

An airplane has a fuselage. The fuselage is where you take your __5__ and your luggage is stowed. There are also windows so you can see mountains, or, depending on your __6__, watch obstacles, such as financial pressure, growth targets, increased automation, competition or staff shortages fade away. To be able to fly, you need __7__ and a tail, but these alone will not get you in the air. Nothing works without large and reliable __8__. The design and functionality of the individual components of an aircraft are a __9__ for your organization and its __10__, and eventually someone has to decide where the journey goes.

| A. profession | B. obstacles | C. metaphor | D. wings | E. effectively |
| F. airplane | G. carefree | H. seat | I. employees | J. engines |

Reading

Explained: The Importance of Human Factors Trainings

A. Safety in a Complex World

In our ever-changing world, filled with numerous **obstacles** and **complexities**, ensuring safety becomes **paramount**. We strive to protect ourselves from known risks, but the reality is that unforeseen challenges can arise at any moment. To navigate through these complexities, we have developed tools that allow us to **overcome** obstacles quickly, effectively, and most importantly, safely. One such tool that **exemplifies** this is the airplane.

B. The Airplane **Metaphor**

An airplane serves as a powerful metaphor for understanding the importance of human factors trainings in organizations. Just as an airplane consists of various **components**, each playing a crucial role, an organization is made up of individuals working together towards a common goal. Let's explore this metaphor in more detail.

C. **Wings** of the Organization

The wings of an airplane are responsible for carrying its weight and enabling it to **soar** through the skies. Similarly, in organizations, the selection, education, and training of employees act as the wings. Properly selecting individuals who **possess** the necessary skills and knowledge, providing them with comprehensive education, and ongoing training are **vital** for the success of the organization. However, just like wings alone cannot get an airplane off the ground, training alone is not **sufficient** to propel an organization forward.

D. Motivation as the Engine

An airplane relies on powerful engines to **generate** the necessary **thrust** to move forward. Similarly, in organizations, employee motivation acts as the engine that **propels** the organization towards its objectives. Motivated employees are driven to **excel** in their roles, **contributing** their best efforts and ideas to achieve success. It is crucial for organizations to foster a supportive environment that encourages and **nurtures** the motivation of its workforce. Proper recognition, rewards, and career growth opportunities are essential fuel for this engine.

E. Communication as Fuel

In an airplane, fuel ensures its continuous operation. Likewise, communication serves as the fuel that runs through every organization. Effective communication is vital for seamless coordination, information sharing, and **collaboration** among team members. Without clear and open lines of communication, organizations may **encounter** issues such as misunderstandings, lack of **alignment**, and missed opportunities. It is **imperative** for organizations to **promote** a culture of effective communication, where information flows freely and **transparently**.

F. Smooth and **Coordinated** Flight

The successful flight of an airplane relies on the smooth coordination of its various components. The **rudder** and **elevator**, controlled from the **flight-deck**, ensure a **stable** and coordinated flight, **compensating** for any **disturbances** and maintaining a desired **trajectory**. Similarly, in organizations, effective teamwork, collaboration, and coordination are crucial for achieving collective goals. Lack of control or poor coordination can lead to **turbulence** within the organization, **hindering** progress and causing discomfort for all involved.

G. Awareness of External Threats

Just as pilots rely on **radar** systems to **detect** and **interpret** external threats, organizations must remain **vigilant** to **anticipate** and **mitigate** potential obstacles. Continuous monitoring and analysis of the external environment, staying informed about industry trends, competitors, and **emerging** risks, are essential for **proactive** decision-making. By keeping a watchful eye on the radar, organizations can identify potential **hazards**, adapt their strategies, and navigate through challenges successfully.

H. **Landing-Gear** for Maximum Stress

The landing gear of an airplane, though not visible during **cruising altitude**, plays a critical role during takeoff and landing. It must **withstand immense** pressure and provide a **cushioning** effect, ensuring a safe and smooth **touchdown**. Similarly, within organizations, human factors trainings equip employees with the necessary knowledge and skills to handle unexpected and complex situations. These trainings prepare individuals to operate effectively under maximum stress, ensuring the organization can navigate through challenging times with **resilience**.

I. Preparedness for **Emergencies**

While we hope for the best, it is essential to be prepared for emergencies. Airplanes are equipped with appropriate emergency equipment, emergency exits, and **evacuation** procedures to ensure the safety of passengers in case of unexpected events. Similarly, organizations should have **robust contingency** plans, crisis management strategies, and training programs in place to handle unforeseen circumstances effectively. Being prepared enhances an organization's ability to respond **swiftly**, maintain operational continuity, and **safeguard** the well-being of its employees and stakeholders.

J. The Role of Effective Teamwork

As we look towards the future, new obstacles will inevitably arise, regardless of the industry or sector. It is unwise to **assume** that everything will work out without **deliberate** effort. Effective teamwork, collaboration, and individual contribution play a **crucial** role in determining the success or failure of organizations, not only in the present but also in the ever-evolving future. By **fostering** a culture that values teamwork, encourages **diverse** perspectives, and promotes a shared vision, organizations can navigate through challenges, adapt to changing circumstances, and thrive in a **dynamic** world.

In summary, human factors are of paramount importance in organizations. By understanding the significance of each component in the airplane metaphor, organizations can cultivate a culture that **prioritizes** employee selection, education, and training, fosters motivation and effective communication, ensures smooth coordination, remains aware of external threats and industry trends.

New Words and Expressions

obstacle *n.* /ˈɑːbstək(ə)l/	sth. that stands in the way 障碍，障碍物
complexity *n.* /kəmˈpleksəti/	the state of being formed of many parts and difficult to understand 复杂性，难懂
paramount *adj.* /ˈpærəmaʊnt/	more important than anything else 最重要的，首要的
overcome *v.* /ˌəʊvərˈkʌm/	to succeed in dealing with or controlling a problem that has been preventing you from achieving sth. 克服
exemplify *v.* /ɪɡˈzemplɪfaɪ/	to give an example in order to make sth. clearer 举例说明，例证，例示
metaphor *n.* /ˈmetəfə(r)/	an expression to refer to sth. that it does not literally denote in order to suggest a similarity 暗喻，隐喻

component *n.* /kəmˈpəʊnənt/	a part of sth.
	部件
wing *n.* /wɪŋ/	a movable organ of birds or planes for flying (one of a pair)
	翅膀，机翼
soar *v.* /sɔː(r)/	to rise quickly and smoothly up into the air
	升空，升腾
possess *v.* /pəˈzes/	to have or own sth.
	拥有
vital *adj.* /ˈvaɪt(ə)l/	urgently needed; absolutely necessary
	至关重要的
sufficient *adj.* /səˈfɪʃnt/	enough for a particular purpose; as much as you need
	足够的，充足的
▲generate *v.* /ˈdʒenəreɪt/	to produce or create sth.
	产生，引起
thrust *n.* /θrʌst/	the force that is produced by an engine to push a plane, rocket, etc. forward
	推力，驱动力
propel *v.* /prəˈpel/	to move, drive or push sth. forward
	推动，驱动，推进
excel *v.* /ɪkˈsel/	to be very good at doing sth.
	擅长，善于，突出
contribute *v.* /kənˈtrɪbjuːt/	to increase, improve or add to sth.
	增加，增进，添加（到某物）
nurture *v.* /ˈnɜːtʃə(r)/	to care for and protect sb./sth. while they are growing and developing
	养育，养护，培养
collaboration *n.* /kəˌlæbəˈreɪʃn/	the act of working with others to create or produce sth.
	合作，协作
encounter *v.* /ɪnˈkaʊntə(r)/	to experience sth., especially sth. unpleasant or difficult, while you are trying to do sth. else

Unit 3 Aviation Humanity

遭遇，遇到（尤指令人不快或困难的事）

alignment *n.* /əˈlaɪnmənt/ — arrangement in a straight line; a position of agreement or alliance
排成直线，协议，结盟

imperative *adj.* /ɪmˈperətɪv/ — very important and needing immediate attention or action
重要紧急的，迫切的，急需处理的

promote *v.* /prəˈməʊt/ — to help sth. to happen or develop
促进，推动

transparently *adv.* /trænsˈpærəntli/ — easily understood or seen through
显然地，易觉察地

coordinated *adj.* /kəʊˈɔːdɪneɪtɪd/ — harmonious, concerted
协调一致的

rudder *n.* /ˈrʌdər/ — a piece of wood or metal at the back of a boat or an aircraft that is used for controlling its direction
（船的）舵，（飞机的）方向舵

elevator *n.* /ˈelɪveɪtər/ — the airfoil on the tailplane of an aircraft that makes it ascend or descend
升降舵

flight-deck *n.* /ˈflaɪt dek/ — an area at the front of a large plane where the pilot sits to use the controls and fly the plane
（飞机的）驾驶舱

stable *adj.* /ˈsteɪbl/ — firmly fixed; not likely to move, change or fail
稳定的，稳固的，牢固的

compensate *v.* /ˈkɒmpenseɪt/ — to provide sth. good to balance or reduce the bad effects of damage, loss, etc.
补偿，弥补

disturbance *n.* /dɪˈstɜːrbəns/ — the act of disturbing something or someone; setting something in motion
干扰

trajectory *n.* /trəˈdʒektəri/ — the curved path of sth. that has been fired,

	hit or thrown into the air
	（射体在空中的）轨道，弹道，轨迹
turbulence *n.* /ˈtɜːbjələns/	a situation with a lot of sudden change, confusion, disagreement and sometimes violence
	骚乱，动乱，动荡，混乱
hinder *v.* /ˈhɪndə(r)/	to make it difficult for sb. to do sth. or sth. to happen
	阻碍，妨碍，阻挡
radar *n.* /ˈreɪdɑːr/	measuring instrument in which the echo of a pulse of microwave radiation is used to detect and locate distant objects
	雷达
detect *v.* /dɪˈtekt/	to discover or notice sth., especially sth. that is not easy to see, hear, etc.
	发现，查明，侦察
interpret *v.* /ɪnˈtɜːrprət/	to explain the meaning of sth.
	诠释，说明
vigilant *adj.* /ˈvɪdʒɪlənt/	very careful to notice any signs of danger or trouble
	警觉的，警惕的，警戒的，谨慎
anticipate *v.* /ænˈtɪsɪpeɪt/	to expect sth.
	预料，预期
mitigate *v.* /ˈmɪtɪɡeɪt/	to make sth. less harmful, serious, etc.
	减轻，缓和
emerging *adj.* /ɪˈmɜːrdʒɪŋ/	new or developing
	新兴的，发展初期的
proactive *adj.* /ˌprəʊˈæktɪv/	controlling a situation by making things happen rather than waiting for things to happen and then reacting to them
	积极主动的，主动出击的，先发制人的
hazard *n.* /ˈhæzərd/	danger; a possibility of causing loss, misfortune or damage
	危险，危害

landing-gear *n.* /ˈlændɪŋ gɪr/ — undercarriage, the part of an aircraft, including the wheels, that supports it when it is landing and taking off
起落架，起落装置，着陆装置

gear *n.* /gɪr/ — a piece of machinery used for a particular purpose
（特定用途的）器械，装置

cruise *v.* /kruːz/ — to travel at a steady speed
以平稳的速度行驶

altitude *n.* /ˈæltɪtuːd/ — the height above sea level
海拔，高程

withstand *v.* /wɪðˈstænd/ — to be strong enough not to be hurt or damaged by extreme conditions, the use of force, etc.
承受，抵住，顶住，经受住

immense *adj.* /ɪˈmens/ — extremely large or great
极大的，巨大的

cushion *v.* /ˈkʊʃn/ — to make the effect of a fall or hit less severe
（跌倒或碰撞时）起缓冲作用

touchdown *n.* /ˈtʌtʃdaʊn/ — the moment when a plane or spacecraft lands
（飞机或宇宙飞船的）着陆，降落，接地

resilience *n.* /rɪˈzɪliəns/ — the ability of people or things to feel better quickly after sth. unpleasant, such as shock, injury, etc.
快速恢复的能力，适应力

emergency *n.* /ɪˈmɜːrdʒənsi/ — a sudden unforeseen crisis (usually involving danger) that requires immediate action
突发事件，紧急情况

evacuate /ɪˈvækjueɪt/ *v.* — to move people from a place of danger to a safer place
（把人从危险的地方）疏散，转移，撤离

robust *adj.* /rəʊˈbʌst/ — (of a system or an organization) strong and not likely to fail or become weak

	强劲的，富有活力的
contingency *n.* /kənˈtɪndʒənsi/	an event that may or may not happen
	可能发生的事，偶发（或不测、意外）事件
swiftly *adv.* /ˈswɪftli/	quickly, rapidly, promptly
	很快地，敏捷地，即刻
safeguard *v.* /ˈseɪfɡɑːd/	to protect sth./sb. from loss, harm or damage; to keep sth./sb. safe
	保护，保障，捍卫
assume *v.* /əˈsuːm/	to think or accept that sth. is true but without having proof of it
	假定，假设，认为
deliberate *adj.* /dɪˈlɪbərət/	done on purpose rather than by accident
	故意的，蓄意的，存心的
crucial *adj.* /ˈkruːʃ(ə)l/	of extreme importance; vital to the resolution of a crisis
	至关重要的
foster *v.* /ˈfɒstə(r)/	to encourage sth. to develop
	促进，助长，培养，鼓励
diverse *adj.* /daɪˈvɜːs/	very different from each other and of various kinds
	不同的，相异的，多种多样的，形形色色的
dynamic *adj.* /daɪˈnæmɪk/	always changing and making progress
	动态的，发展变化的
prioritize *v.* /praɪˈɒrətaɪz/	to put tasks, problems, etc. in order of importance, so that you can deal with the most important first; to treat sth. as being more important than other things
	按重要性排列，优先处理

Genre Analysis

Genre:
The passage can be classified as an informative/explanatory text, specifically focusing on the importance of human factors trainings in organizations. It uses a metaphorical approach by likening organizational dynamics to the various components of an airplane, making the

information more relatable and engaging.

Communicative Moves:

Move 1 Introduction to the Topic (A):

Establishes the context by highlighting the importance of safety in a complex world, setting the stage for the subsequent discussion on human factors training.

Move 2 Metaphorical Introduction (B):

Introduces the airplane metaphor, connecting the components of an airplane to the organizational structure to provide a clear analogy for understanding the importance of human factors training.

Move 3 Analogies and Comparisons (C, D, E, F, G, H, I):

Each subsection elaborates on a specific aspect of the metaphor, drawing parallels between the components of an airplane and various elements within organizations, such as employee selection, motivation, communication, coordination, awareness of external threats, preparedness for emergencies, and handling stress.

Move 4 Emphasizing Key Points (C, D, E, F, G, H, I):

Reinforces the importance of employee selection, motivation, communication, coordination, awareness, preparedness, and stress management in organizational success through the metaphorical lens of an airplane.

Move 5 Future Outlook and Recommendations (J):

Looks towards the future and emphasizes the role of effective teamwork and collaboration in navigating challenges and ensuring organizational success in an evolving environment.

Move 6 Conclusion (J):

Summarizes the key points discussed throughout the passage, underlining the importance of human factors training in organizations and how understanding the airplane metaphor can help in fostering a conducive organizational culture.

Overall, the passage effectively communicates the significance of human factors trainings through the use of vivid analogies, clear comparisons, and a structured approach that guides the reader through various aspects of organizational functioning.

Communicative Purpose:

1. Inform and Educate: It explains why human factors training is crucial for organizations.
   Persuade and Advocate: It convinces readers to invest in such training for organizational success.
2. Illustrate Concepts: It uses the airplane analogy to make complex ideas easy to understand.
3. Empower Decision-making: By showing the benefits of training, it encourages leaders to

prioritize it.

4. Inspire Action: It urges readers to foster teamwork and readiness for future challenges.

Culture Tips

1. Prioritize employee selection and development: Place importance on selecting individuals with the right skills and values for your organization. Provide comprehensive education and ongoing training to continuously develop their knowledge and abilities.

2. Foster motivation and recognition: Create a supportive environment that encourages and nurtures employee motivation. Recognize and reward employees for their contributions, and provide opportunities for career growth and advancement.

3. Promote effective communication: Establish a culture of open and transparent communication. Encourage information sharing, active listening, and clear and concise messaging to ensure seamless coordination and collaboration among team members.

4. Encourage teamwork and collaboration: Foster a collaborative work environment where teams can work together towards common goals. Promote the sharing of diverse perspectives, encourage cooperation, and provide opportunities for team-building activities.

5. Stay vigilant and adapt to external factors: Continuously monitor and analyze the external environment, including industry trends and emerging risks. Foster a mindset of adaptability and proactivity to anticipate challenges and make informed decisions.

6. Embrace a learning culture: Encourage a culture of continuous learning and improvement. Provide opportunities for professional development, training, and knowledge sharing to enhance employee skills and capabilities.

7. Promote resilience and preparedness: Develop robust contingency plans and crisis management strategies to handle unforeseen circumstances. Foster a culture of preparedness and equip employees with the necessary knowledge and skills to handle challenging situations.

8. Value effective leadership: Cultivate effective leadership at all levels of the organization. Encourage leaders to provide guidance, support, and clear direction to their teams, promoting a sense of stability and trust.

9. Foster a shared vision: Create a shared vision that aligns employees towards common goals and objectives. Encourage employees to connect with the organization's mission and values, fostering a sense of purpose and collective identity.

10. Embrace innovation and adaptability: Encourage a culture of innovation, where employees are empowered to generate new ideas and approaches. Emphasize

Unit 3 Aviation Humanity

adaptability and agility to respond to changing circumstances and market demands.

11. By incorporating these culture tips, organizations can create an environment that values employee development, effective communication, teamwork, adaptability, and resilience, ultimately fostering success in a complex world.

Reading Exercises

I. Vocabulary Matching

Directions: *Fill in the blank with the corresponding letter of the Chinese equivalent of the following English words.*

1. phrase   (　)　8. component   (　)　15. reliable   (　)　22. compensate   (　)
2. obstacle   (　)　9. eventually   (　)　16. generate   (　)　23. disturbance (　)
3. issue   (　)　10. proportional (　)　17. freeze   (　)　24. coordination (　)
4. overcome   (　)　11. selection   (　)　18. explode   (　)　25. detect   (　)
5. luggage   (　)　12. education   (　)　19. run out   (　)
6. automation   (　)　13. vital   (　)　20. distribute   (　)
7. profession   (　)　14. extremely   (　)　21. schedule   (　)

A. 可靠的　　　　H. 组件　　　　　O. 短语　　　　V. 补偿
B. 障碍　　　　　I. 最终　　　　　P. 生成　　　　W. 干扰
C. 问题　　　　　J. 成比例的　　　Q. 冻结　　　　X. 耗尽
D. 克服　　　　　K. 选择　　　　　R. 爆炸　　　　Y. 检测
E. 极为　　　　　L. 教育　　　　　S. 协调
F. 自动化　　　　M. 至关重要的　　T. 日程
G. 职业　　　　　N. 行李　　　　　U. 分发

II. Answer the Following Questions

1. What role does motivation play in organizations, according to the passage?
2. Why is effective communication crucial for organizations?
3. How does the passage emphasize the importance of teamwork and coordination in organizations?
4. What is the significance of human factors trainings in organizations?
5. Why is it important for organizations to stay aware of external threats and industry trends?

III. Multiple Choice

1. What is the metaphor used to emphasize the importance of human factors training in organizations?　(　)

A. Ship.　　　B. Train.　　　C. Airplane.　　　D. Car.

2. Which component in the airplane metaphor represents the selection, education, and training of employees in organizations? (　)

   A. Wings.　　　B. Engines.　　　C. Rudder.　　　D. Landing gear.

3. What serves as the fuel that runs through every organization, according to the passage? (　)

   A. Wings.　　　B. Engines.　　　C. Communication.　　　D. Radar.

4. What is the role of effective teamwork and coordination in organizations? (　)

   A. To generate motivation.　　　B. To ensure smooth flights.
   C. To monitor external threats.　　　D. To achieve collective goals.

5. Why is it important for organizations to have robust contingency plans and crisis management strategies? (　)

   A. To select and educate employees effectively.
   B. To ensure smooth flights.
   C. To handle unforeseen circumstances.
   D. To foster effective communication.

Translation

Directions: *Translate the following sentences into Chinese.*

1. We have developed tools to overcome obstacles quickly, effectively and safely.

   _____
   _____

2. There are also windows so you can see mountains, or, depending on your profession, watch obstacles, such as financial pressure, growth targets, increased automation, competition or staff shortages fade away.

   _____
   _____

3. The wings of your organization are the selection, education, and training of your employees, but the wings alone will not get the plane off the ground.

   _____
   _____

4. An engine is extremely reliable as long as it's taken care of.

   _____
   _____

5. The fact is nothing runs without fuel.

6. Lack of control or poor coordination can lead to a turbulent flight even without external factors.

7. Good pilots keep their eye on the radar and think far enough ahead to identify hazards and time to avoid them.

8. Within a company, this vital gear is all of the necessary training for unexpected and complex situations when an organization is exposed to maximum stress.

9. The airplane is equipped with appropriate emergency equipment, emergency exits and slides to evacuate every passenger safely and quickly.

10. The organization, each individual, and in particular effective teamwork perform a crucial role, as they determine success or failure, not only today, but above all in the future.

Speaking

Student: Why are human factors trainings important for organizations?
Aviation expert: Human factors trainings are important because they help organizations ensure safety and navigate through complexities. They are like the different components of an airplane that work together for a successful flight.
Student: How do the wings of an airplane relate to employee selection, education, and training?
Aviation expert: Just as wings carry an airplane's weight and enable it to fly, selecting

the right employees and providing them with education and training is crucial for an organization's success.

Student: What role does employee motivation play in organizations?

Aviation expert: Employee motivation acts as the engine that propels an organization towards its objectives. Motivated employees contribute their best efforts and ideas for success.

Student: How does effective communication contribute to organizations?

Aviation expert: Effective communication is like the fuel that runs through an organization. It enables seamless coordination, information sharing, and collaboration among team members.

Student: Why is teamwork and coordination important for organizations?

Aviation expert: Just as smooth coordination is necessary for a successful flight, effective teamwork and coordination are crucial for organizations to achieve their goals and function smoothly.

Student: How does being aware of external threats relate to organizational success?

Aviation expert: Organizations must stay vigilant about external factors that can impact their success. By monitoring industry trends, competitors, and risks, they can adapt their strategies and make informed decisions.

Student: How does effective teamwork and collaboration help organizations navigate challenges?

Aviation expert: Effective teamwork and collaboration enable organizations to leverage the collective knowledge and skills of their employees. This fosters innovation, problem-solving, and adaptability, helping them overcome challenges and thrive in a dynamic environment.

Student: Thank you for simplifying it. Now I understand why human factors trainings are important for organizations.

Writing

Directions: *For this part, you are allowed 30 minutes to write a composition entitled "My Thoughts on the University Arts Festival". You should write at least 120 words following the outline given below in Chinese:*

1. 表达你对即将举行的大学艺术节的看法；

2. 对艺术节具体内容和组成部分的建议。

_____
_____
_____
_____
_____
_____
_____
_____

Expansion Exercise for CET-BAND 4—Reading

Directions: *The passage is followed by some questions or unfinished statements. For each of them there are four choices marked A, B, C and D. You should decide on the best choice.*

In the coming era of budget cuts to education, distance learning could become the norm.

The temptation for those in charge of education budgets to trade teachers for technology could be so strong that they ignore the disadvantages of distance learning. School facilities are expensive to build and maintain, and teachers are expensive to employ. Online classes do not require buildings and each class can host hundreds of people simultaneously, resulting in greater savings, thus increasing the temptation of distance education for those concerned more about budgets than learning. But moving away from a traditional classroom in which a living, breathing human being teaches and interacts with students daily would be a disaster. Physically attending school has hidden benefits: getting up every morning, interacting with peers, and building relationships with teachers are essential skills to cultivate in young people. Moreover, schools should be more than simple institutions of traditional learning. They are now places that provide meals. They are places where students receive counseling and other support.

Those policy-makers are often fascinated by the latest technology in education and its potential to "transform" education overnight. But online education does not allow a teacher to keep a struggling student after class and offer help. Educational videos may deliver academic content, but they are unable to make eye contact or assess a student's level of engagement. Distance education will never match the personal teaching in a traditional classroom. In their first 18 years of life, American children spend only 9% of their time in school. Yet teachers are expected to prepare them to be responsible citizens, cultivate their social skills, encourage successful time management, and enhance their capacity to flourish in an increasingly harsh labor market. Given these expectations, schools should not become permanently "remote".

The power of the classroom is rooted in the humanity of the people gathered in the same place, at the same time. Personal teaching is about teachers showing students a higher path, and about young people going through the process together. Technology, no matter how advanced, should simply be a tool of a good teacher.

1. What mainly accounts for the possibility that distance learning could become the norm? ( )

   A. Advances in education technology.
   B. Shrinking financial resources.
   C. Shortage of school facilities.
   D. Lack of qualified teachers.

2. What does the author say is one possible benefit of students attending school physically? ( )

   A. Developing the habit of getting up early.
   B. Eating nutritionally well-balanced meals.
   C. Growing into living and breathing human beings.
   D. Cultivating relationships with peers and teachers.

3. What does the author think of the latest technology in education? ( )

   A. It may have potential disadvantages.
   B. It may render many teachers jobless.
   C. It may add to students' financial burden.
   D. It may revolutionize classroom teaching.

4. What does the author say teachers are expected to do? ( )

   A. Enhance students' leadership capacity.
   B. Elevate students to managerial positions.
   C. Enable students to adapt to the changes in life.
   D. Prepare students to be competitive in the future.

5. Why couldn't technology replace a good teacher? ( )

   A. It lacks humanity.   B. It is still immature.
   C. It cannot track students' growth.   D. It cannot cater to personal needs.

Unit 4 Aviation Education

Preview

The passage provides a list of various job opportunities within the aviation industry, specifically focusing on jobs in airlines. It highlights that there are diverse roles beyond flight and cabin crews, encompassing positions in HR, IT, sales, marketing, and project management. The passage then provides short descriptions of several airline jobs. The passage emphasizes the diversity of career options within the aviation industry and provides a glimpse into the roles, qualifications, and salaries associated with each position.

航空基础英语

Listening

Directions: *Fill in the blanks with the words that you hear. Find them in the table below and write down the corresponding letters.*

There are all kinds of job roles within the airlines. Most people tend to think no further than the flight and __1__ __2__, but of course there are all the roles you would expect in any large company, from HR to IT, from sales and marketing to project management.

Avionics technicians are responsible for the __3__ and maintenance of avionics equipment in aircraft. They must be able to read __4__ and __5__, as well as have a working __6__ of __7__ and computers.

Airline administrative support personnels are the __8__ of the airline industry. They keep the airline running __9__ and efficiently behind the scenes. This important role is responsible for a wide range of activities, from answering customer inquiries to preparing flight __10__. Airline administrative support personnels are essential to a successful airline operation.

A. knowledge	B. installation	C. crew	D. smoothly	E. paperwork
F. blueprint	G. electronics	H. backbone	I. schematic diagrams	J. cabin

Reading

Aviation Jobs in Airlines and Airports

When it comes to aviation jobs, the sky is not the limit, as the old saying goes "the sky is home". Here is a list with short descriptions of many opportunities for employment within the aviation industry. Which of these attracts you?

There are all kinds of job roles within the airlines. Most people tend to think no further than the flight and **cabin** crews, but of course there are all the roles you would expect in any large company, from HR to IT, from sales and marketing to project management.

Aircraft Maintenance Technician

An aircraft maintenance technician

is responsible for inspecting, maintaining and repairing the various parts and systems that make up an aircraft. They work on everything from the engines and wings to the landing gear and tail. The job can be challenging, but it's also very rewarding when everything is working properly and the aircraft takes off safely. Aircraft maintenance technicians typically have a high school **diploma** or **equivalent**, although some may have an **associate's degree** in aviation maintenance technology. They can expect to earn a **salary** of around $50,000 per year.

Airline Administrator

Airline administrative support personnels are the **backbone** of the airline industry. They keep the airline running smoothly and efficiently behind the scenes. This important role is responsible for a wide range of activities, from answering customer inquiries to preparing flight paperwork. Airline administrative support personnels are essential to a successful airline operation. If you are interested in a career in aviation, airline administrative support may be a good fit for you. This **position** requires excellent customer service skills as well as strong organizational and problem-solving abilities. It is a challenging but rewarding career, and there are many opportunities for advancement within the airline industry.

Airline Attorney

Airline attorneys are responsible for ensuring that an airline is in **compliance** with all federal and state laws. They also provide legal advice to the airline's management and employees, represent the airline in **civil litigation** and negotiate in **draft contracts**. Airline attorneys must have a law degree from an **accredited** law school, be licensed to practice law in the state where the airline is based, and have several years of experience in aviation law. They should also be familiar with **the UK Civil Aviation Federal Aviation Administration**'s (FAA) regulations governing airlines. Airline attorneys typically work for large airlines or law firms that specialize in aviation law. They may work full-time or part-time and may be required to travel.

Airline Manager

Airline managers are responsible for the overall operation of an airline. They develop and implement policies and **procedures**, direct and manage **staff** and **oversee** financial operations. To be successful in this role, it is important to have strong leadership skills and be able to make decisions quickly. Airline managers typically have a college degree in business or aviation, and they can expect to earn a salary of around $85,000 per year.

Airline Passenger Assistant

Airline passenger assistants are responsible for helping passengers with boarding and deplaning as well as providing general assistance to passengers. Passenger assistants must be able to work quickly and efficiently and they need to have excellent people skills. They also need to be able to handle difficult situations calmly and effectively. A college degree is not typically required for this position, but it is helpful to have some customer service or aviation experience.

Airline Pilot

Airline pilots are responsible for the safe transportation of passengers and **cargo**. They must ensure that all safety procedures are followed, and they must be able to handle any emergency that may arise. In addition, airline pilots must be able to **navigate** their aircraft through all weather conditions. They must also be able to communicate effectively with air traffic control and with the **cabin crew**. Airline pilots can earn very high salaries later in their careers. However, the salaries can be surprisingly low for junior first officers. It all depends on the airline. In addition, pilots can receive excellent benefits, including health **insurance** and generous **retirement package**s.

Airline Ramp Agent

Airline ramp agents are responsible for the safety and security of passengers and cargo on an aircraft. They also help to load and unload baggage and cargo as well as direct planes on the ground. This is a very important role in the aviation industry, and airline ramp agents must be able to work quickly and efficiently to get planes boarded and off the ground safely. A college degree is not typically required for this position, but it is helpful to have some experience in customer service or aviation.

Airline **Regional** Sales Manager

Airline regional sales managers are responsible for developing and maintaining relationships with key **clients** in their region. They work with airline **executives** to create sales plans and identify new growth opportunities. They must be able to build strong relationships with clients and be able to negotiate favorable contracts. If you have a **passion** for aviation and want a career that will keep you on your toes, consider becoming an airline regional sales manager.

Airline Reservation Sales Agent

Airline reservation sales agents are the front line of customer service for airlines. They are responsible for helping passengers book flights, **reserve** seats, and make changes to their reservations. They also handle customer **complaints** and provide assistance as needed. The role of airline reservation sales agent can be a challenging but rewarding one. It is a great opportunity for young people who are interested in travel and customer service.

Airline Ticket Agent

Airline ticket agents are responsible for a wide variety of tasks, such as issuing tickets, checking in passengers and handling baggage. They must be able to work quickly and efficiently, and they need to have excellent customer service skills. A college degree is not typically required for this position, but it is helpful to have some customer service or aviation experience. Airline ticket agents can expect to earn a salary of around $35,000 per year.

Avionics Technician

Avionics technicians are responsible for the installation and maintenance of avionics equipment in aircraft. They must be able to read blueprints and **schematic diagrams**, as well as have a working knowledge of electronics and computers. Avionics technicians typically work for airlines, aircraft **manufacturers**, or maintenance companies. The role of an avionics technician is to maintain and install **communication, navigation, and surveillance equipment** on aircraft. They must be able to read blueprints and schematic diagrams as well as have a good understanding of electronics and computers. Avionics technicians typically work for airlines, aircraft manufacturers or maintenance companies.

Crew Schedule Coordinator

A crew schedule **coordinator** is responsible for creating, maintaining, and managing the crew schedules for an airline. They work with the pilots, flight **attendants**, and other airline staff to ensure that all flights are staffed with the necessary personnel, and that all crew members are properly rested and meet all regulatory requirements. This is a **critical** role in ensuring that flights operate safely and on time.

Flight Attendant

Flight attendants play a crucial role in ensuring the safety of passengers onboard an aircraft. They are responsible for performing a wide variety of tasks, from providing customer service to overseeing the safety of the cabin. To be a successful flight attendant, it is important to have strong customer service skills and be able to handle stress effectively. The median **annual** salary for flight attendants is $50,500. However, salaries can vary significantly, depending on the airline company and the level of experience. **Entry-level** flight attendants typically earn around $35,000 per year, while those with more experience can earn up to $75,000 per year. Most airlines offer benefits packages that include medical and dental insurance, paid vacation and 401(k) plans. Some airlines also offer sign-on bonuses and other incentives.

Flight Instructor

Airline flight **instructors** are responsible for inducting recently **qualified** pilots into the airline, and ensuring that pilots maintain the high standards of flying required. They have to ensure that the pilots are sufficiently skilled to fly the various aircraft types, and that they have all the recent requirements for them. Flight instructors are often **recruited** from the most senior pilots who have **demonstrated** their abilities during many years of flying.

Human Resources Manager

As a human resources manager for an airline, you would be responsible for overseeing the **hiring** process, managing employee files and benefits, and ensuring compliance with labor laws. This is a challenging and rewarding career that requires excellent organizational skills, attention to detail, and the ability to handle difficult conversations.

Information Technology Specialist

Airline information technology specialists are responsible for managing the technology that keeps airline operations running smoothly. They work with computers and software to **track** flights, monitor baggage and cargo, and manage reservations and other customer data. This is a critical role, because without a strong and stable IT **infrastructure**, the airline would be **vulnerable** to not only **flight delay**s and loss of business but also **cyber-attacks**.

New Words and Expressions

▲cabin *n.* /ˈkæbɪn/ — small room on a ship or boat where people sleep
机舱，客舱

▲maintenance *n.* /ˈmeɪntənəns/ — activity involved in maintaining something in good working order
维护，保养

diploma *n.* /dɪˈpləʊmə/ — a document certifying the successful completion of a course of study
毕业文凭，学位证书

equivalent *adj.* /ɪˈkwɪvələnt/ — equal in amount or value
等同的

Unit 4 Aviation Education

▲**salary** *n.* /ˈsæləri/	money that employees receive for doing their job, especially professional employees or people working in an office, usually paid every month	
	薪金，薪水（尤指按月发放的）	
backbone *n.* /ˈbækbəʊn/	a central cohesive source of support and stability	
	支柱，骨干	
▲**position** *n.* /pəˈzɪʃ(ə)n/	a job in an organization	
	职务，职位	
▲**compliance** *n.* /kəmˈplaɪəns/	acting according to certain accepted standards	
	服从，遵守	
accredited *adj.* /əˈkredɪtɪd/	given official approval to act	
	官方认可的	
▲**procedure** *n.* /prəˈsiːdʒər/	a process or series of acts especially of a practical or mechanical nature involved in a particular form of work	
	程序	
▲**staff** *n.* /stæf/	personnel who assist their superior in carrying out an assigned task	
	全体员工	
oversee *v.* /ˌəʊvərˈsiː/	watch and direct	
	监管，监督	
▲**cargo** *n.* /ˈkɑːrɡəʊ/	goods carried by a large vehicle	
	（船或飞机装载的）货物	
navigate *v.* /ˈnævɪɡeɪt/	act as the navigator in a car, plane, or vessel and plan, direct, plot the path and position of the conveyance	
	导航，引路	
▲**insurance** *n.* /ɪnˈʃʊrəns/	protection against future loss	
	保险	
▲**regional** *adj.* /ˈriːdʒən(ə)l/	of a certain place	
	地区的，地域的	

▲ **client** *n.* /ˈklaɪənt/ someone who pays for goods or services
客户，委托人

executive *n.* /ɪɡˈzekjətɪv/ a person responsible for the administration of a business
主管，经理

passion *n.* /ˈpæʃ(ə)n/ strong feeling or emotion
酷爱，热衷的爱好（或活动等）

reserve *v.* /rɪˈzɜːrv/ obtain or arrange (for oneself) in advance
预订

complaint *n.* /kəmˈpleɪnt/ an expression of grievance or resentment
抱怨，投诉

▲ **manufacturer** *n.* /ˌmænjuˈfæktʃərər/ a business engaged in manufacturing some product
生产商，制造商

▲ **coordinator** *n.* /kəʊˈɔːrdɪneɪtər/ someone whose task is to see if the work goes harmoniously
协调人，统筹者

▲ **attendant** *n.* /əˈtendənt/ someone who waits on or tends to or attends to the needs of another
服务人员

▲ **critical** *adj.* /ˈkrɪtɪk(ə)l/ urgently needed; absolutely necessary
极其重要的，关键的

annual *adj.* /ˈænjuəl/ occurring or payable every year
年度的

entry-level *adj.* primary or junior
初级的

instructor *n.* /ɪnˈstrʌktər/ a person whose occupation is teaching
教练，导师

qualified *adj.* /ˈkwɑːlɪfaɪd/ holding appropriate documentation and officially on record as qualified to perform a specified function or practice a specified skill
具备……的学历（资历）的

recruit *v.* /rɪˈkruːt/ seek to employ

		招聘，招收
demonstrate *v.* /ˈdemənstreɪt/		establish the validity of something, as by an example, explanation or experiment
		证明
hire *v.* /ˈhaɪər/		engage or hire for work
		聘用，录用
track *v.* /træk/		go after with the intent to catch
		跟踪（进展情况）
infrastructure *n.* /ˈɪnfrəstrʌktʃər/		the basic structure or features of a system or organization
		基础设施
vulnerable *adj.* /ˈvʌlnərəb(ə)l/		capable of being wounded or hurt
		易受伤害的
associate's degree		副学士学位
civil litigation		民事诉讼
draft contract		合同草案
the UK Civil Aviation Federal Aviation Administration		英国民航联邦航空局
law firms		律师事务所
cabin crew		机组人员
retirement package		退休金
schematic diagram		示意图
communication navigation and surveillance equipment		通信、导航和监视设备
flight delay		航班延误
cyber-attack		网络攻击

Genre Analysis

Genre:

The passage can be categorized as an informative and persuasive piece within the domain of career or job exploration. It provides a comprehensive list of various aviation jobs available in airlines, along with brief descriptions and qualifications, aiming to inform readers about the wide range of opportunities within the aviation industry and potentially attract their interest.

Communicative Moves:

Move 1 Introducing the Topic:

The passage begins with a catchy phrase and an assertion that aviation jobs go beyond the limits of the sky, setting the tone for the exploration of diverse career options.

Move 2 Listing and Describing Job Roles:

The passage presents a series of aviation job roles within airlines, providing concise descriptions for each position. It highlights the responsibilities, qualifications, and sometimes the salary expectations associated with each job.

Move 3 Highlighting Benefits and Opportunities:

The passage emphasizes the rewarding aspects of the mentioned job roles, such as high salaries, career advancement prospects, benefits packages, and incentives. This serves to entice potential candidates and spark their interest.

Move 4 Providing Additional Information:

In some cases, the passage offers supplementary details about specific job requirements, industry regulations, or the importance of certain roles. This adds depth to the descriptions and helps readers understand the significance of each position.

Move 5 Concluding Remarks:

The passage closes with a note on the critical role of information technology specialists in maintaining a robust IT infrastructure within airlines, highlighting the potential consequences of neglecting this aspect.

Communicative Purpose:

The passage aims to inform readers about the diverse array of aviation job opportunities available within airlines. It provides a brief overview of each job role, including responsibilities and qualifications, with the intention of attracting individuals who may be interested in pursuing a career in aviation. By emphasizing the rewards, benefits, and advancement possibilities associated with these roles, the passage seeks to persuade readers to consider exploring and potentially applying for these positions.

Culture Tips

1. Embrace teamwork: The aviation industry thrives on teamwork and collaboration. Cultivate strong teamwork skills and be prepared to work closely with colleagues from various departments to ensure safe and efficient operations.

2. Prioritize professionalism: Professionalism is highly valued in the aviation industry. Maintain a high level of professionalism in your interactions with colleagues, superiors,

and customers. Demonstrate integrity, reliability, and a strong work ethic.

3. Commitment to safety: Safety is of paramount importance in aviation. Develop a strong commitment to safety protocols and procedures. Strive to maintain a safety-first mindset in all aspects of your work.

4. Adaptability and flexibility: The aviation industry is dynamic and ever-changing. Be adaptable and flexible in your approach to work. Embrace new technologies, industry advancements, and changing circumstances with an open mindset.

5. Attention to detail: Attention to detail is crucial in aviation, where even the smallest oversight can have significant consequences. Develop a keen eye for detail and cultivate a meticulous approach to your work.

6. Effective communication: Effective communication is essential in the aviation industry. Practice clear, concise, and respectful communication with colleagues, superiors, and customers. Be an active listener and ensure that information is conveyed accurately and efficiently.

7. Respect for diversity: The aviation industry is diverse, with employees from different backgrounds, cultures, and nationalities. Embrace and respect this diversity. Cultivate an inclusive and welcoming environment that celebrates differences and fosters mutual respect.

8. Customer focus: Airlines rely on providing excellent customer service. Develop strong customer service skills and maintain a customer-centric approach in your interactions. Strive to meet and exceed customer expectations to ensure a positive travel experience.

9. Continuous learning: The aviation industry is constantly evolving. Cultivate a mindset of continuous learning and professional development. Stay updated on industry trends, advancements in technology, and regulatory changes to enhance your skills and knowledge.

10. Passion for aviation: Show genuine passion for the aviation industry. Let your enthusiasm shine through your work and interactions. Cultivate a deep appreciation for the industry's unique challenges, rewards, and contributions to global connectivity.

11. By keeping these cultural tips in mind, you can better adapt to the aviation industry's work environment, values, and expectations, setting yourself up for success in your chosen career path.

Reading Exercises

I. Vocabulary Matching

Directions: *Fill in the blank with the corresponding letter of the Chinese equivalent of the* *following English words.*

1. airline ( )   7. flight ( )   13. safety ( )   19. experience ( )
2. airport ( )   8. passengers ( )   14. airway ( )   20. baggage ( )
3. aviation ( )   9. customer ( )   15. crew ( )   21. sales ( )
4. responsible ( )   10. role ( )   16. ensuring ( )   22. agents ( )
5. work ( )   11. skills ( )   17. service ( )
6. aircraft ( )   12. pilots ( )   18. typically ( )

A. 飞行   G. 航空公司   M. 安全   S. 经验
B. 机场   H. 乘客   N. 空中航线   T. 行李
C. 航空   I. 客户   O. 机组人员   U. 销售
D. 通常   J. 代理   P. 确保   V. 角色
E. 工作   K. 技能   Q. 服务
F. 飞机   L. 飞行员   R. 负责任的

II. Answer the Following Questions

1. What are other job roles in aviation apart from flight and cabin crews?
2. What are the qualifications and responsibilities of aircraft maintenance technicians?
3. What are the main responsibilities of airline pilots?
4. What tasks do airline ramp agents perform?
5. What skills are important for flight attendants, and what salary and benefits do they receive?

III. Multiple Choice

1. Which aviation job requires a law degree and expertise in aviation law? ( )

   A. Airline administrator
   B. Airline pilot.
   C. Airline attorney.
   D. Aircraft maintenance technician.

2. Which aviation job involves installing and maintaining avionics equipment in aircraft?
   ( )

   A. Flight attendant.
   B. Avionics technician.
   C. Airline ramp agent.
   D. Crew schedule coordinator.

3. What is the role of a crew schedule coordinator in the aviation industry? ( )

   A. Ensuring the safety and security of passengers and cargo.

   B. Developing and maintaining relationships with key clients.

   C. Creating, maintaining, and managing crew schedules.

   D. Overseeing the hiring process and managing employee files.

4. What is the main responsibility of an airline information technology specialist? ( )

   A. Overseeing the hiring process and managing employee files.

   B. Managing the technology that keeps airline operations running smoothly.

   C. Inspecting, maintaining, and repairing various parts of an aircraft.

   D. Assisting passengers with boarding and deplaning.

5. Which aviation job involves overseeing the overall operation of an airline and implementing policies and procedures? ( )

   A. Airline Manager.

   B. Airline Reservation Sales Agent.

   C. Airline Passenger Assistant.

   D. Airline Ticket Agent.

Translation

Directions: *Translate the following sentences into Chinese.*

1. When it comes to aviation jobs, the sky is not the limit, as the old saying goes "the sky is home".

2. An aircraft maintenance technician is responsible for inspecting, maintaining and repairing the various parts and systems that make up an aircraft.

3. If you are interested in a career in aviation, airline administrative support may be a good fit for you.

4. Airline attorneys must have a law degree from an accredited law school, be licensed to

practice law in the state where the airline is based, and have several years of experience in aviation law.

5. Pilots must ensure that all safety procedures are followed, and they must be able to handle any emergency that may arise.

6. Airline sales representatives also handle customer complaints and provide assistance as needed.

7. The role of an avionics technician is to maintain and install communication, navigation, and surveillance equipment on aircraft.

8. To be a successful flight attendant, it is important to have strong customer service skills and be able to handle stress effectively.

9. Flight attendants play a crucial role in ensuring the safety of passengers on board an aircraft.

10. Flight instructors are often recruited from the most senior pilots who have demonstrated their abilities during many years of flying.

11. Unless the staff members are wearing the uniform of a particular airline, all the people you see working at an airport are employed by the airport itself, employed by one of the companies with a concession at the airport, or their contractors.

12. It is also a great way to get started in the aviation industry, as mechanics can often move

Unit 4 Aviation Education

up to more specialized positions with experience.

_____
_____

13. The role of a flight dispatcher is essential to the safety of passengers and crew. They are responsible for the coordination of all ground operations related to a flight including loading and unloading baggage and cargo, fueling the aircraft and arranging catering.

_____
_____

Speaking

Student: Hello, I recently came across some information about aviation jobs within airlines, and it got me thinking about pursuing a career in the aviation industry. I'd love to get some guidance from a career counselor regarding the available opportunities.

Career counselor: That's wonderful to hear! The aviation industry offers a wide range of exciting career paths. I'd be happy to assist you. What specific information are you seeking?

Student: I'm particularly interested in learning more about the different job roles within airlines. The passage mentioned a few, but I'd like to explore them further and understand which one might be the best fit for me.

Career counselor: Of course! Let's dive into it. Besides the obvious roles like pilots and cabin crew, there are various other positions within airlines. Some examples include aircraft maintenance technicians, air traffic controllers, customer service representatives, aviation engineers, and even roles within the HR department, such as recruitment, employee relations, and performance management.

Student: That's quite diverse! Could you provide more details about some of these roles? I'm especially curious about aircraft maintenance technicians and aviation engineers.

Career counselor: Certainly! Aircraft maintenance technicians are responsible for inspecting, repairing, and maintaining aircraft to ensure they meet safety standards. They play a vital role in keeping the planes in top condition. On the other hand, aviation engineers focus on designing and developing aircraft systems, structures, and components. They work on improving efficiency, safety, and performance of aircraft.

Student: Those sound fascinating! What qualifications or educational background do these roles typically require?

Career counselor: For aircraft maintenance technicians, a diploma or degree in aircraft maintenance or a related field is usually required. You'll also need to obtain the necessary certifications and licenses. For aviation engineers, a bachelor's degree in aerospace or mechanical engineering is typically required. Specialized knowledge in areas like aerodynamics and aircraft systems is highly valued.

Student: I see. What about career growth and advancement opportunities in these roles?

Career counselor: Both roles offer opportunities for growth and advancement. With experience and additional certifications, aircraft maintenance technicians can progress to supervisory or management positions. Aviation engineers can specialize in specific areas or take on higher-level design and research roles. Continuous learning, staying updated on industry advancements, and gaining practical experience are key to advancing in these careers.

Student: That's good to know. Lastly, are there any specific skills or qualities that would make someone successful in these roles?

Career counselor: Absolutely! Attention to detail, problem-solving abilities, technical aptitude, and a strong commitment to safety are crucial for both roles. Good communication skills, teamwork, and the ability to work well under pressure are highly valued in the aviation industry. Being adaptable and having a passion for aviation are also important qualities to thrive in these careers.

Student: Thank you for all this valuable information! It's given me a lot to consider. I appreciate your guidance.

Career counselor: You're very welcome! I'm glad I could help. Remember, exploring your interests and passions is key to finding a fulfilling career. If you have any more questions or need further assistance, feel free to reach out. Good luck on your journey towards a career in the aviation industry!

Writing

Directions: *Suppose you have taken part in a career planning seminar. Write an account of the seminar and summarize what you have learned from it. You will have 30 minutes for this task and should write at least 120 words but no more than 180 words.*

_____

_____

Unit 4 Aviation Education

_____
_____
_____
_____
_____
_____

Expansion Exercise for CET-BAND 4—Reading

Directions: *In this section, there is a passage with ten blanks. You are required to select one word for each blank from a list of choices given in a word bank following the passage. Choose the corresponding letter for each blank.*

Public perception of success in the U.S. might be totally misguided.

While 92% of people believe others care most about fame and __1__, fewer than 10% factor those qualities into their own success. This is according to the newly __2__ study by Harvard Graduate School of Education professor Todd Smith. Smith says he was __3__ by how past studies on success "assumed what people will care about". In this study, his team "went the __4__ direction" by spending years carrying out individual interviews and group surveys to see what people really talk about when they talk about success.

As a scientist, Smith __5__ studied individuality for a living, and even he was surprised to find younger respondents cared more about having a __6__ in life. Those between the ages of 18 and 34 prioritized it most, and that prioritization dropped off as respondents' ages went up. Perhaps this is because older people had fewer options when they were starting their careers, at a time when values focused more on stable incomes than __7__ personal missions.

Other trends included an emphasis on the importance of parenting. Being a parent __8__ very high across the priorities of all study participants. Ultimately, Smith hopes institutions will take note of these insights __9__.

Higher education institutions tend to focus on preparing students for high-paying jobs. For such institutions, from universities to workplaces, to better __10__ people in the U.S., "They'll need to understand what the American public highly prioritizes," Smith says.

A. accommodate B. accordingly C. acquiring D. bothered E. fortune
F. fulfilling G. identify H. literally I. opposite J. profession
K. purpose L. ranked M. released N. similarly O. wrong

Unit 5 Aviation Legend

Preview

Feng Ru, a Chinese immigrant to the United States, became fascinated with aviation and sought to bring it back to his native country. After working in various industries to gain mechanical knowledge, Feng turned his attention to aviation upon learning about the Wright brothers' success. He established the Guangdong Air Vehicle Company in Oakland, California, and built his own airplanes. Despite setbacks and a fire in his workshop, Feng successfully conducted test flights and gained recognition from the press and revolutionary Sun Yat-sen. In 1911, Feng returned to China with a biplane of his own construction to make exhibition flights. However, he faced anti-Chinese sentiment in the United States at the time. Tragically, Feng died in a plane crash during an aerial exhibition in 1912. He was honored with a military funeral and recognized as a pioneer of Chinese aviation.

Listening

Directions: *Fill in the blanks with the words that you hear. Find them in the table below and write down the corresponding letters.*

Feng became well known for developing __1__ versions of the water pump, the __2__, the telephone, and the __3__ telegraph, some of which were used by San Francisco's Chinese businessmen. But upon hearing of the Wright brothers' success, Feng turned his attention to aviation, __4__ translating into Chinese anything he could find on the Wrights, Glenn Curtiss and, later, French aircraft designer Henri Farman.

By 1906, Feng decided to return to California to __5__ an aircraft factory, building airplanes of his own design. San Francisco's __6__ earthquake and resulting fire forced him to __7__ to Oakland instead, where, __8__ by local Chinese businessmen, Feng __9__ his workshop—a ten-by-eight-foot shack. __10__ into this small space were tools, books, journals, mechanical projects, aircraft parts—and Feng himself, who rarely finished work before 3 a.m.

| A. alternate | B. establish | C. funded | D. Jammed | E. generator |
| F. wireless | G. relocate | H. laboriously | I. massive | J. erected |

Reading

Feng Ru—the Father of Chinese Aviation

A. Feng Ru made history on the California coast, then introduced airplanes to his native land.

B. At **twilight** on a Tuesday evening in September 1909, Feng Ru prepared to test an airplane of his own design above the gently rolling hills of Oakland, California. It was just 6 years after Orville and Wilbur Wright took to the skies at Kill Devil Hills, North Carolina, and only a year after their first public flights.

C. Feng immigrated to the U.S. from China sometime between 1894 and 1898, when he was in his early teens, and immediately set to work doing **odd-jobs** in San Francisco. "He was amazed by America's power and **prosperity**. He understood that **industrialization** made the country great, and felt that industrialization could do the same for China," says historian Patti Gully, who has co-authored a book on the contributions of Chinese living outside their country to the development of aviation in China. "So he went east to learn all he could about machines, working in **shipyards**, **power plants**, machine shops, anywhere he could **acquire** mechanical knowledge."

D. Feng became well known for developing **alternate versions** of the water pump, the **generator**, the telephone, and the wireless telegraph, some of which were used by San Francisco's Chinese businessmen. But upon hearing of the Wright brothers' success, Feng turned his attention to aviation, **laboriously** translating into Chinese anything he could find on the Wrights, Glenn Curtiss and, later, French aircraft designer Henri Farman.

E. By 1906, Feng decided to return to California to establish an aircraft factory, building airplanes of his own design. San Francisco's **massive** earthquake and resulting fire forced him to **relocate** to Oakland instead, where, **funded** by local Chinese businessmen, Feng **erected** his workshop—a ten-by-eight-foot **shack**. **Jammed** into this small space were tools, books, journals, mechanical projects, aircraft parts—and Feng himself, who rarely finished work before 3 a.m.

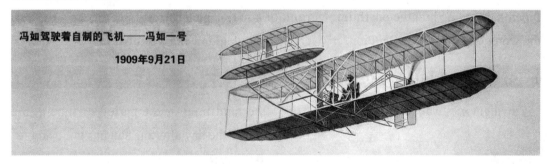

F. In this tiny spot, the self-taught engineer established the Guangdong Air Vehicle Company in 1909, and completed his first airplane that year, according to the **American Institute of Aeronautics and Astronautics**. During one test flight, Feng lost control of his airplane (not an unusual **occurrence**), which **plunged** into his workshop, setting it **ablaze**. Feng and his three assistants moved operations to an Oakland **hayfield**, referred to by the New York Times and the Washington Post as "a hidden **retreat**".

G. To keep his invention secret, he had the engine **castings** made by different East Coast machine shops, then **assembled** the parts himself. His **discretion paid off**: Feng's successful test flights were covered by the **mainstream press**, and his work was **praised** by revolutionary Sun Yat-sen. By 1911, as the New York Times reported on February 21st "[Feng] will leave here for his native land tomorrow, taking with him a biplane of the Curtiss type, in which he intends to make **exhibition** flights. It is believed that he will be the first aviator to rise from the ground in China. The machine he is taking to China is of his own construction. The aviator is **financed** by six of his **countrymen**, **residents** of Oakland, who will **accompany** him on the trip. The first flights will be tried at Hong Kong and Canton."

H. Feng was leaving just in time: anti-Chinese **sentiment** was on the rise in the American West, and the Oregonian reported of the pilot's latest flight "Immigration officials and **customs** inspectors are today said to be **gnashing** their teeth. They find it hard enough to keep the Chinese out now, without having them **dropping in on** flying machines."

I. Feng arrived in Hong Kong on March 21, 1911. On August 26, 1912, Feng was killed while performing an **aerial exhibition**. As he lay dying, he reportedly told his **assistants**, "Your **faith** in the progress of your **cause** is **by no means** to be **affected** by my death."

J. The Republic of China gave Feng Ru a full military funeral, awarding him the **posthumous rank** of a **major general.** At Sun Yat-sen's request, the words "Chinese Aviation Pioneer" were **engraved** upon Feng's **tombstone**.

New Words and Expressions

twilight *n.* /ˈtwaɪlaɪt/	the faint light or the period of time at the end of the day after the sun has gone down
	暮色，薄暮
	（船或飞机的）航向，航线
prosperity *n.* /prɒˈsperəti/	the state of being successful
	兴旺，繁荣
industrialization *n.* /ɪnˌdʌstriələˈzeɪʃn/	the development of industry on an extensive scale
	工业化
shipyard *n.* /ˈʃɪpjɑːd/	a place where ships are built or repaired
	船坞，造船厂，修船厂
▲acquire *v.* /əˈkwaɪə(r)/	to gain sth. through efforts, ability or behaviour
	（通过努力、能力、行为表现）获得，得到
alternate *adj.* /ˈɔːltərnət/	alternative
	可供替代的

Unit 5 Aviation Legend

version *n.* /ˈvɜːrʒn/	a form of sth. that is slightly different from an earlier form or from other forms of the same thing
	变体，变种，型式
generator *n.* /ˈdʒenəreɪtər/	a machine for producing electricity
	发电机
laboriously *adv.* /ləˈbɔːriəsli/	hard, painfully
	辛苦地，费力地
massive *adj.* /ˈmæsɪv/	extremely large or serious
	巨大的，非常严重的
relocate *v.* /ˌriːləʊˈkeɪt/	to move sb./sth. to a new place
	（使）搬迁，迁移
fund *v.* /fʌnd/	to provide money for sth., usually sth. official
	为……提供资金，拨款
erect *v.* /ɪˈrekt/	to build
	建造，建立
shack *n.* /ʃæk/	a small building, usually made of wood or metal, that has not been built well
	简陋的小屋，棚屋
jam *v.* /dʒæm/	to push sth. somewhere with a lot of force
	使劲（往某处）挤（或压、塞）
occurrence *n.* /əˈkʌrəns/	sth. that happens or exists
	发生的事情，存在的事物
plunge *v.* /plʌndʒ/	to move or make sb./sth. move suddenly forwards and/or downwards
	使突然前冲（或下落）
ablaze *adj.* /əˈbleɪz/	burning quickly and strongly
	猛烈燃烧
hayfield *n.* /ˈheɪˌfiːld/	干草地，秣草地
retreat *n.* /rɪˈtriːt/	a quiet, private place that you go to
	僻静处，隐居处
casting *n.* /ˈkɑːstɪŋ/	an object made by pouring hot liquid metal, etc. into a mould (= a specially shaped

77

	container)
	铸件
assemble *v.* /əˈsembl/	to fit together all the separate parts of sth.
	装配，组装
discretion *n.* /dɪˈskreʃn/	care in what you say or do, in order to keep sth. secret or to avoid causing embarrassment to or difficulty for sb.; the quality of being careful
	谨慎，慎重，审慎
mainstream *n.* /ˈmeɪnstriːm/	the ideas and opinions that are thought to be normal because they are shared by most people; the people whose ideas and opinions are most accepted
	主流思想，主流群体
press *n.* /pres/	newspapers and magazines
	报刊杂志，印刷媒体
▲praise *v.* /preɪz/	to express your approval or admiration for sb./sth.
	表扬，赞扬，称赞
exhibition /ˌeksɪˈbɪʃn/ *n.*	the act of showing sth. to the public
	展览，展出
finance *v.* /faɪˈnæns/	to provide money for a project, fund
	提供资金
countryman *n.* /ˈkʌntrimən/	a person born in or living in the same country as sb. else
	同国人，同胞，同乡
resident *n.* /ˈrezɪdənt/	a person who lives in a particular place or who has their home there
	居民，住户
accompany *v.* /əˈkʌmpəni/	to travel or go somewhere with sb.
	陪同，陪伴
sentiment *n.* /ˈsentɪmənt/	a feeling or an opinion, esp. one based on emotions
	（基于情感的）看法，情绪

customs *n.* /ˈkʌstəmz/	海关
inspector *n.* /ɪnˈspektə(r)/	a person whose job is to check that rules are being obeyed and that standards are acceptable 检查员，视察员，巡视员
gnash *v.* /næʃ/	to feel very angry and upset about sth. （气得）咬牙切齿
assistant *n.* /əˈsɪstənt/	a person who helps or supports sb., usually in their job 助理，助手
faith *n.* /feɪθ/	trust in sb.'s ability or knowledge; trust that sb./sth. will do what has been promised 信任，信念，信心
cause *n.* /kɔːz/	an organization or idea that people support or fight for （支持或为之奋斗的）事业，目标，思想
affect *v.* /əˈfekt/	to produce a change in sb. / sth. 影响
posthumous *adj.* /ˈpɒstjʊməs/	happen, done, published, etc. after a person has died 死后发生的
rank *n.* /ræŋk/	the position that sb. has in the army, navy, police, etc. 军衔，军阶，警衔
engrave *v.* /ɪnˈɡreɪv/	to cut words or designs on wood, stone, metal, etc. 在……上雕刻（字或图案）
tombstone *n.* /ˈtuːmstəʊn/	a stone that is used to mark a grave 墓碑
odd job	零工
power plant	发电厂
American Institute of Aeronautics and Astronautics	美国航空航天学会
pay off	还清，得到好结果，获得汇报，取得成功

drop in	顺道拜访，非正式访问
aerial exhibition	航空展览（会）
by no means	绝不
major general	少将

Genre Analysis

Genre:

Historical narrative

Communicative Moves:

Move 1 Providing historical background and context. (Paragraph C)

Move 2 Introducing the main protagonist and his achievements. (Paragraphs A, B)

Move 3 Describing the challenges and dedication of the protagonist. (Paragraphs C, D, E)

Move 4 Highlighting the secrecy and recognition of the protagonist's work. (Paragraph G)

Move 5 Discussing the protagonist's departure from the United States. (Paragraph H)

Move 6 Depicting the protagonist's tragic death and his final words. (Paragraph I)

Move 7 Concluding with the recognition and honors received by the protagonist. (Paragraph J)

Communicative Purpose:

The purpose of this text is to provide a historical account of Feng Ru, his contributions to aviation, and his significance as a pioneer in Chinese aviation. It aims to showcase Feng Ru's determination, innovation, and the recognition he received for his accomplishments despite facing challenges and adversity. The text emphasizes his role as a trailblazer and highlights his impact on the development of aviation in China.

Culture Tips:

1. Determination and perseverance: Feng Ru's story reflects the importance of determination and perseverance in achieving one's goals. Despite facing challenges and setbacks, Feng Ru continued to pursue his passion for aviation and worked tirelessly to develop his own airplanes.

2. Embracing industrialization: Feng Ru recognized the power and prosperity brought about by industrialization in America. This cultural tip highlights the significance of embracing technological advancements and industrial development for societal progress.

3. Value of knowledge and learning: Feng Ru actively sought opportunities to acquire mechanical knowledge and skills. His dedication to learning from various sources exemplifies the Chinese cultural value of valuing education and continuous learning.

Unit 5 Aviation Legend

4. Entrepreneurship and resourcefulness: Feng Ru established his own aircraft factory, even in limited space, and sought financial support from local Chinese businessmen. This demonstrates the entrepreneurial spirit and resourcefulness often valued in Chinese culture.

5. Secrecy and discretion: Feng Ru took measures to keep his inventions and work secret by having engine castings made by different machine shops. This indicates a cultural inclination towards discretion and protecting intellectual property.

6. Patriotism and national pride: Feng Ru's desire to bring aviation to China and contribute to his homeland's development reflects a sense of patriotism and national pride. This cultural value emphasizes the importance of serving one's country and contributing to its progress.

7. Honor and recognition: Feng Ru was given a full military funeral and posthumously awarded the rank of a major general, with the words "Chinese Aviation Pioneer" engraved on his tombstone. This highlights the cultural significance of honor, recognition, and paying tribute to individuals who have made significant contributions to their country.

Reading Exercises

I. Vocabulary Matching

Directions: *Fill in the blank with the corresponding letter of the Chinese equivalent of the following English words.*

1. revolutionary ( ) 8. wireless ( ) 15. plunge ( )
2. exhibition ( ) 9. military ( ) 16. pioneer ( )
3. mainstream ( ) 10. assemble ( ) 17. rank ( )
4. mechanical ( ) 11. workshop ( ) 18. aerial ( )
5. occurrence ( ) 12. funeral ( ) 19. retreat ( )
6. inspector ( ) 13. version ( ) 20. vehicle ( )
7. anywhere ( ) 14. general ( )

A. 车辆　　　　　　H. 版本　　　　　　O. 检查员
B. 撤退　　　　　　I. 葬礼　　　　　　P. 发生
C. 空中的　　　　　J. 工作坊　　　　　Q. 机械的
D. 等级　　　　　　K. 组装　　　　　　R. 主流
E. 先锋　　　　　　L. 军事的　　　　　S. 展览
F. 跳水　　　　　　M. 无线的　　　　　T. 革命的
G. 将军 / 一般的　　N. 任何地方

II. Answer the Following Questions

1. Who was Feng Ru and what did he do?

2. Where did Feng Ru test his airplane?

3. Why did Feng Ru return to California in 1906?

4. How did Feng Ru keep his invention a secret?

5. How was Feng Ru honored after his death?

III. Multiple Choice

1. When did Feng Ru first get his airplane into the sky? ( )

    A. A year after the Orville and Wilbur Wright brothers' first public flight.

    B. A year before the Orville and Wilbur Wright brothers' first public flights.

    C. When he introduced airplanes to his native land.

    D. On a Tuesday evening in September 1910.

2. What is true about Feng Ru according to Patti Gully? ( )

    A. He did all he could to acquire medical knowledge.

    B. He wanted to make China great through industrialization.

    C. Feng immigrated to the U.S. from China in the late 18th century.

    D. Feng went back to China immediately.

3. Which of the following was NOT mentioned in the paragraph as an achievement by Feng Ru? ( )

    A. Water pump.

    B. The generator.

    C. The automobile.

    D. The wireless telegraph.

4. What is NOT true about Feng Ru's aircraft factory—Guangdong Air Vehicle Company? ( )

    A. It was established in California.

    B. It was established in Oakland.

    C. It was funded by local Chinese businessmen.

    D. It was burned accidentally during one test flight.

5. How was Feng Ru covered by the New York Times in 1911? ( )

    A. The aviator is financed by four of his countrymen.

    B. The machine he was about to take back to China was not of his own construction.

    C. He would be the first aviator in China.

    D. His first exhibition flight would be just in Hong Kong.

Translation

Directions: *Translate the following sentences into Chinese.*

1. Feng Ru was amazed by America's power and prosperity. He understood that industrialization made the country great, and felt that industrialization could do the same for China.

2. But upon hearing of the Wright brothers' success, Feng turned his attention to aviation, laboriously translating into Chinese anything he could find on the Wrights, Glenn Curtiss and, later, French aircraft designer Henri Farman.

3. By 1906, Feng decided to return to California to establish an aircraft factory, building airplanes of his own design.

4. In this tiny spot, the self-taught engineer established the Guangdong Air Vehicle Company in 1909, and completed his first airplane that year, according to the American Institute of Aeronautics and Astronautics.

5. Feng's discretion paid off: his successful test flights were covered by the mainstream press, and his work was praised by revolutionary Sun Yat-sen.

6. As Feng lay dying, he reportedly told his assistants, "Your faith in the progress of your cause is by no means to be affected by my death."

Speaking

Chinese student: Have you heard about Feng Ru, the Father of Chinese Aviation?

Foreigner: No, I haven't. Tell me more about him.

Chinese student: Feng Ru was a remarkable figure in the history of aviation. He made history in California and then introduced airplanes to China. He was born in China but immigrated to the United States in his early teens.

Foreigner: That's fascinating! What inspired him to pursue aviation?

Chinese student: Feng Ru was amazed by America's power and prosperity. He believed that industrialization could do the same for China. After working odd jobs in San Francisco, he gained mechanical knowledge and became well-known for developing various inventions.

Foreigner: How did he become involved in aviation?

Chinese student: When he heard about the Wright brothers' success, Feng Ru turned his attention to aviation. He studied their work, as well as that of Glenn Curtiss and Henri Farman. Despite facing setbacks, he established the Guangdong Air Vehicle Company in 1909 and built his first airplane.

Foreigner: Building an airplane must have been quite a challenge. How did he manage it?

Chinese student: Feng Ru faced many challenges, but his determination was strong. He worked tirelessly in a small workshop, even after his first airplane crashed into it. To keep his invention a secret, he had different parts made by machine shops across the East Coast and assembled them himself.

Foreigner: That's impressive. Did he receive recognition for his work?

Chinese student: Yes, he did. Feng Ru's successful test flights gained attention from the mainstream press, and even Sun Yat-sen, a prominent revolutionary figure, praised his work. In 1911, he left for China with his own biplane, becoming the first aviator to rise from the ground in China.

Foreigner: It's amazing how he managed to achieve so much. What happened after he returned to China?

Chinese student: Unfortunately, shortly after his return, Feng Ru was killed during an aerial exhibition in 1912. Despite his tragic death, he left a lasting legacy. The Republic of China recognized his contributions by giving him a full military funeral and engraving "Chinese Aviation Pioneer" on his tombstone.

Foreigner: Feng Ru's story is truly inspiring. He played a crucial role in the development of aviation in China.

Chinese student: Absolutely! His dedication, innovation, and recognition as a pioneer have left a significant mark in Chinese aviation history.

Foreigner: Thank you for sharing this remarkable story with me. It's incredible to learn about individuals like Feng Ru who made such significant contributions.

Chinese student: You're welcome! It's always inspiring to delve into the stories of innovators and pioneers like Feng Ru, who shape the world we live in today.

Writing

Directions: *For this part you are allowed 30 minutes to write an essay on your choice of friends. Some people choose friends who are different from them. Others choose friends who are similar to them. Which kind of friend would you choose? You should write at least 120 words but no more than 180 words.*

Expansion Exercise for CET-BAND 4—Reading

Directions: *In this section, there is a passage with ten blanks. You are required to select one word for each blank from a list of choices given in a word bank following the passage. Choose the corresponding letter for each blank.*

  It is commonly believed that the great English dramatist and poet William Shakespeare was born in Stratford-on-Avon on April 23, 1564. But it is impossible to know the __1__ day on which he was born.

  Church records show he was baptized (施洗礼) on April 26, and three days was a customary amount of time to wait before baptizing a newly born baby. Shakespeare's date of death is __2__ known, however: it was April 23, 1616. He was 52 and had retired to Stratford three years before.

Although few plays have been performed or analyzed as extensively as the 38 plays Shakespeare wrote, there are few surviving details about his life. This __3__ of biographical information is due primarily to his social __4__; he was not a noble, but the son of a leather trader.

Shakespeare __5__ attended the grammar school in Stratford, where he would have studied Latin and read __6__ literature. He did not go to university and at age 18 married Anne Hathaway, who was eight years his __7__. They had four children, including the twins, Hamnet and Judith. Nothing is known of the period between the birth of the twins and Shakespeare's __8__ as a dramatist in London in the early 1590s.

In a million words written over 20 years, he __9__ the full range of human emotions and conflicts with a __10__ that remains sharp today. As his great contemporary the poet and dramatist Ben Jonson said "He was not of an age, but for all time."

A. captured	B. classical	C. conclusively	D. emergence	E. exact
F. generated	G. particular	H. position	I. precision	J. probably
K. quality	L. scarcity	M. senior	N. separated	O. systematically

Unit 6 Aviation Feat

Preview

On January 15, 2009, US Airways Flight 1549, piloted by Captain Chesley "Sully" Sullenberger and First Officer Jeffrey Skiles, experienced a bird strike shortly after takeoff from La Guardia Airport in New York. Both engines lost power, leaving the plane in a critical situation. Realizing they couldn't make it back to an airport, Sullenberger made the decision to land the plane in the Hudson River. With skillful maneuvering, the plane successfully landed on the water, and all 155 passengers and crew members were evacuated safely before the plane sank. The event came to be known as the "Miracle on the Hudson" and garnered international attention.

Sullenberger and Skiles were hailed as heroes, and the successful ditching maneuver was recognized as one of the most remarkable in aviation history. The story was later adapted into a movie titled "Sully: Miracle on the Hudson" released in 2016.

Listening

Directions: *Fill in the blanks with the words that you hear. Find them in the table below and write down the corresponding letters.*

On January 15, 2009, US Airways Flight 1549 departed from La Guardia Airport in New York. The captain, Chelsey Sullenberger, a seasoned pilot with a background in aviation safety, was accompanied by First Officer Jeffrey Skiles. The flight seemed routine, carrying 150 passengers and three cabin crew members, until the unexpected happened. __1__, the plane encountered a flock of Canada geese, resulting in a collision that caused both engines to fail.

With no engine power and limited options, Captain Sullenberger swiftly took control of the situation. He attempted to restart the engines but to no avail. Recognizing the impossibility of reaching a runway safely, he made the bold decision to land the plane in the Hudson River. __2__, he knew the river was their only chance for survival. The aircraft descended rapidly, and at 3:31 PM, it made a controlled water landing.

Miraculously, all passengers and crew members survived the impact. __3__ the chaos, Captain Sullenberger's quick thinking and calm demeanor played a vital role in the successful outcome. The cabin crew promptly initiated the evacuation, ensuring the safety of everyone on board. The frigid Hudson River presented new challenges as water flooded the aircraft. Despite the chaos, the rescue operation swiftly began.

Within minutes, rescue boats arrived to assist in the evacuation. __4__, passengers were directed to inflatable rafts and the wings of the partially submerged plane. The authorities had been alerted, and the Coast Guard was prepared to provide aid. Thanks to the efforts of the rescue teams, all passengers and crew members were safely evacuated from the plane.

Captain Sullenberger's remarkable decision-making and the subsequent successful water landing earned him widespread recognition. In recognition of his remarkable __5__, he and his crew were honored with Master's Medals, a prestigious award for their exceptional achievements in aviation. The individuals involved in the rescue operation also received Certificates of Honor, acknowledging their efforts.

The events of US Airways Flight 1549 captured the attention of people worldwide. The __6__, it was a testament to the courage and competence of those involved. In September 2016, the story was immortalized in the movie *Sully: Miracle on the Hudson* further showcasing the remarkable events of that fateful day.

The __7__ and bravery of the crew and the resilience of the passengers remain an inspiration. This incident serves as a reminder that even in the face of unforeseen challenges, human ingenuity and determination can prevail. They deserve our utmost __8__ for their ability to make split-second decisions that saved lives. The "Miracle on the Hudson" will forever be remembered as a remarkable example of the triumph of human spirit and skill.

In the aftermath of the incident, investigations revealed that returning to La Guardia Airport would have been impossible. __9__, Captain Sullenberger's prompt decision to land in the Hudson River proved to be the only viable option. The ditching maneuver executed by the captain was hailed as the most successful in aviation history.

The incredible story of US Airways Flight 1549 serves as a testament to the unwavering dedication and professionalism of the aviation industry. __10__ the plane landed safely in the Hudson River, sparing the lives of all on board. The courage and resourcefulness displayed on that day continue to inspire and remind us of the extraordinary capabilities of humanity.

A. Consequently B. Ultimately   C. Despite   D. respect   E. Thus
F. Realizing   G. Suddenly   H. resilience   I. heroism   J. achievement

Reading

Miracle on the Hudson

A. On January 15, 2009, US Airways Flight 1549 was a routine flight from La Guardia Airport in New York to Charlotte Douglas Airport in North Carolina. The captain of the Airbus A-320 was 57-year-old Chelsey Sullenberger (nicknamed "Sully"), an expert on aviation safety and a former glider pilot. Before joining civil aviation in 1980, he'd been a fighter pilot and served in the United States Air Force. By this fateful day, Sullenberger had already **log**ged over 19,000 flight hours, including almost 5,000 in an A320. The First Officer was 49-year-old Jeffrey Skiles, with over 15,000 flight hours, although none of them were on an Airbus A320. Besides the two pilots, there were 150 passengers and three cabin crew members on board the plane.

B. At 3:24 PM, the pilots were all set and ready, and the La Guardia Airport control tower cleared the flight for takeoff. All the **equipment** worked without a flaw, and the weather was great, with 10-mile (16 km) **visibility** that provided the cockpit crew with a breathtaking view of the Hudson.

C. And then, abruptly, it happened. At 3:27 PM, with the plane moving at 316 ft (96 m) per second, captain Sullenberger noticed a flock of Canada geese just a moment before the plane struck the birds. Moving at such a high speed at the altitude of 2,800 ft (about 854 m), the pilots could not avoid the **collision**. They could only stare, helplessly, at the large birds that filled their view. Then, the aircraft shuddered, and everybody inside heard rumbling sounds and deafening bangs coming from the engines. The passengers sitting in the window seats saw flames coming out of the engines and then ... there was only a deafening silence and the gut-wrenching smell of **fuel**. Every person on the plane must have been thinking at that moment, "It can't be happening to me!"

D. However, Sully didn't lose his presence of mind, even after he realized that both engines had shut down. He understood there was no chance that the aircraft would end up somewhere on a runway, undamaged. So, just seconds after the collision, Sullenberger sprang into action. He took over control while Skiles was trying to restart the engines. Unfortunately, his first attempt didn't bring any result. But then, the co-pilot managed to at least turn on the **auxiliary power unit** (APU); without which, the aircraft would have been uncontrollable. The pilots wouldn't be able to go left or right, up or down, and the plane would be literally falling out of the sky. With the APU on, the plane was still climbing, but it was visibly slowing down. After reaching a height of 3,060 ft (930 m), the aircraft started a gliding descent while speeding up to 240 miles (390 kph) per hour. At 3:28 PM, it had already descended 1,650 ft (500 m). The situation was dire and required immediate actions. Some 20 seconds after the plane lost its **engine power**, Captain Sullenberger sent out a **Mayday distress** call. "Hit the birds, **no thrust in both engines**, back to La Guardia," was his desperate report. Air traffic controller Patrick Harten ordered the airport tower to **halt** all **departures** and told Sullenberger to head for Runway 13. But the captain responded with only one word, "Unable".

E. By that time, the experienced pilot had already realized that the plane wouldn't be able to make it back to the airport. He told Harten that they might end up in the Hudson. The air traffic controller couldn't believe his ears and didn't want to accept the terrifying truth. He tried to find a way to get the plane to the runway, but at 3:29 PM, Sullenberger repeated his words, only this time, he sounded even more sure: the plane WAS going to end up in the river. Even though Sullenberger had never performed such a maneuver before, he was ready to sacrifice the plane to save the lives of the passengers and crew. The only place in the densely populated New York Metropolitan area that would be long, wide, and smooth enough to land a heavy plane descending at breakneck speed was the Hudson River. After realizing this, the captain spent the next three seconds (an extravagant amount of time in those circumstances) to make his only announcement. Sounding confident and calm, he said, "This is the Captain. Brace for impact." Immediately

after this, he heard the cabin crew shouting commands to the passengers, telling them to brace and stay down. The pilots couldn't pay attention to anything but the descent.

F. The plane flew less than 900 ft (270 m) over the George Washington Bridge while the co-pilot was calling out the airspeed and altitude of the plane. The cockpit was filled with warnings repeated by a computerized voice, "Terrain, too low, too low, caution terrain, pull up, pull up, pull up." And then, at 3:31 PM, the plane skidded into the river. The impact was hard, but the plane was afloat and miraculously intact. Almost in unison, the pilots exclaimed, "It wasn't as bad as I had thought!"

G. But naturally, that wasn't the end of the story—the crew had to evacuate people from a plane that was floating in the middle of the river! Once the plane stopped on the river's surface, the cabin crew started the evacuation. Some of the passengers escaped to the plane's **wing**s; others jumped onto rafts. Earlier, the air traffic controllers had contacted the Coast Guard and asked them to prepare for assisting with the rescue. That's why just four minutes after the river landing, the first rescue boats began to arrive. But the situation was far from safe. One panicked passenger opened one of the **rear doors**, and the flight attendant couldn't manage to **seal** it again. Water started to fill the plane through this door as well as through a hole in the fuselage. The water level was rising fast, and the cabin crew was urging the passengers to climb over the seats and move forward.

H. The last person was evacuated at 3:55 PM. All the passengers and crew members were alive; there were only 5 serious injuries, and 78 people received some sort of treatment, mostly for hypothermia and minor injuries.

I. The partially submerged plane

was later brought to a pier in Lower Manhattan, 4 miles (6 km) from the place where they'd landed. When the authorities investigated afterward, they tested whether it would have been possible for the plane to return to La Guardia airport safely. The answer was no. What's more, it turned out that even a 35-second delay would have killed all the people on board the plane and even more on the ground in the attempt.

J. Also, the **ditching maneuver** Sullenberger had performed was called the most successful in the history of aviation. On January 22, 2009, all the aircraft's crew members got Master's Medals, an award which is given very rarely and only for outstanding aviation achievements. Those who took part in the **rescue operation** were awarded as well; they got Certificates of Honor. In September 2016, people all over the world went to cinemas to watch the Warner Bros. movie "Sully: Miracle on the Hudson," based on the events of that unique flight.

New Words and Expressions

log *n.* /lɒg/	record officially in writing or on a computer 记录，积累（指飞行小时数的记录）
equipment *n.* /ɪˈkwɪpmənt/	things used for a particular purpose 设备
visibility *n.* /ˌvɪzəˈbɪlɪti/	how far or how clearly you can see in particular weather conditions 能见度
collision *n.* /kəˈlɪʒən/	a moving object crashes into sth. 碰撞
fuel *n.* /ˈfjuːəl/	coal, oil, or petrol that is burned to provide heat or power 燃料
halt *v.* /hɔːlt/	stop completely 停止
departure *n.* /dɪˈpɑːrtʃər/	the act of going away from somewhere 起飞
wing *n.* /wɪŋ/	one of the large flat parts that stick out from the side of a plane and help to keep it in the air when it is flying 机翼

seal *v.* /siːl/	provide a tight and perfect closure 关闭
engine power	发动机动力
auxiliary power unit (APU)	辅助动力装置
Mayday distress call	危险求救呼叫
no thrust in both engines	双发失效
rear doors	后舱门
ditching maneuver	水上迫降
rescue operation	救援行动

Genre Analysis

Genre:
The text appears to be a narrative or descriptive account of the events surrounding the incident known as "Miracle on the Hudson". It combines elements of a news report and storytelling.

Communicative moves:

Move 1 Background Information:

The text provides background information about the flight, the pilots, and the number of passengers and crew on board.

Move 2 Incident Description:

The text describes the collision with the birds, the loss of engine power, and the actions taken by the pilots during the crisis.

Move 3 Communication Exchanges:

The text includes exchanges between the pilots and the air traffic controller, conveying the urgency and decision-making process.

Move 4 Dramatic Tension:

The text builds suspense and tension as it describes the descent towards the Hudson River, the warnings from the cockpit, and the final landing.

Move 5 Rescue and Aftermath:

The text describes the evacuation and rescue efforts, as well as the recognition and awards received by the crew and rescue teams.

Move 6 Cultural Impact:

The text mentions the subsequent movie adaptation, highlighting the global interest and recognition of the event.

Communicative Purpose:

The purpose of the text is to inform and recount the extraordinary event of the emergency landing on the Hudson River. It aims to engage the reader through a dramatic narrative and evoke a sense of awe and admiration for the actions taken by the pilots and the successful outcome of the incident.

Culture Tips

1. Preparedness and training: The incident underscores the importance of thorough training and preparedness in any field. Just as pilots undergo extensive training to handle emergencies, it's essential for individuals and organizations to invest in continuous learning and preparation to handle unexpected situations effectively.

2. Teamwork and communication: The successful outcome of the incident was largely due to the effective teamwork and communication among the crew members. Encouraging a culture of open communication, collaboration, and mutual support within teams can greatly enhance problem-solving and decision-making during critical moments.

3. Leadership and quick thinking: Captain Sullenberger's decisive leadership and quick thinking were instrumental in the successful outcome. Cultivating leadership skills and encouraging proactive thinking can empower individuals to make sound decisions even under pressure.

4. Safety culture: The incident serves as a reminder of the paramount importance of safety in any industry. Promoting a safety culture that prioritizes adherence to protocols, reporting of potential risks, and continuous improvement can prevent accidents and mitigate their impact.

5. Resilience and adaptability: The passengers and crew of Flight 1549 demonstrated resilience and adaptability when faced with a life-threatening situation. Cultivating these traits can help individuals and organizations navigate challenges and bounce back from setbacks.

6. Recognition and appreciation: The recognition and awards received by Captain Sullenberger and the crew highlight the importance of acknowledging and appreciating exceptional achievements. Celebrating success and recognizing outstanding performance can boost morale, inspire others, and foster a culture of excellence.

Unit 6 Aviation Feat

Reading Exercises

I. Vocabulary Matching

Directions: *Fill in the blank with the corresponding letter of the Chinese equivalent of the following English words.*

1. log ( )
2. cabin crew ( )
3. jet ( )
4. fuel ( )
5. engine power ( )
6. equipment ( )
7. visibility ( )
8. in-flight tasks ( )
9. collision ( )
10. auxiliary power unit (APU) ( )
11. Mayday distress call ( )
12. no thrust in both engines ( )
13. halt ( )
14. departure ( )
15. rear doors ( )
16. seal ( )
17. rescue operation ( )
18. ditching maneuver ( )
19. wing ( )
20. aboard ( )
21. airline operations manager ( )

A. 机舱乘务人员
B. 喷气式飞机
C. 航空公司运营经理
D. 记录，积累（指飞行小时数的记录）
E. 飞行中任务
F. 密封
G. 辅助动力装置（APU）
H. 危险求救呼叫
I. 能见度
J. 水上迫降动作
K. 机翼
L. 起飞
M. 燃料
N. 发动机动力
O. 双发失效
P. 在船/飞机上
Q. 救援行动
R. 后舱门
S. 设备
T. 碰撞
U. 停止

II. Answer the Following Questions

1. Who were the pilots of US Airways Flight 1549?
2. What caused the emergency situation during the flight?
3. How did Captain Sullenberger handle the situation after the engines shut down?
4. Why did Captain Sullenberger decide to land the plane in the Hudson River?

5. What were the outcomes of the emergency landing in the Hudson River?

III. Multiple Choice

1. What was the reason behind the loss of engine power in US Airways Flight 1549? ( )

    A. A collision with a flock of birds.

    B. Poor weather conditions.

    C. A technical failure in the engines.

    D. An error made by the pilots.

2. How did Captain Sullenberger react after realizing that both engines had shut down? ( )

    A. He immediately called for help from air traffic control.

    B. He prepared the passengers for an emergency landing.

    C. He attempted to restart the engines.

    D. He tried to maneuver the plane back to La Guardia Airport.

3. Why did Captain Sullenberger decide to land the plane in the Hudson River? ( )

    A. The air traffic controller insisted on landing at the airport.

    B. The passengers preferred a water landing over a runway landing.

    C. The plane had lost control and couldn't land safely anywhere else.

    D. The plane couldn't make it back to the airport.

4. What was the outcome of the water landing in the Hudson River? ( )

    A. There were several casualties and injuries.

    B. All passengers and crew members survived.

    C. The plane sank immediately after landing.

    D. The plane exploded upon impact.

5. How was Captain Sullenberger recognized for his actions in the US Airways Flight 1549 incident? ( )

    A. He received an award for outstanding aviation achievements.

    B. He was praised by air traffic control for his quick thinking.

    C. He was given a promotion to a higher rank in the airline.

    D. He became a subject of a Hollywood movie based on the incident.

Translation

Directions: *Translate the following sentences.*

1. The captain of the Airbus A-320 was 57-year-old Chelsey Sullenberger (nicknamed

"Sully"), an expert on aviation safety and a former glider pilot.

2. By this fateful day, Sullenberger had already logged over 19,000 flight hours, including almost 5,000 in an A320.

3. The cockpit crew spent this time gaining altitude and performing a series of necessary in-flight tasks.

4. Moving at such a high speed at the altitude of 2,800 ft (about 854 m), the pilots could not avoid the collision.

5. Then, the aircraft shuddered, and everybody inside heard rumbling sounds and deafening bangs coming from the engines.

6. 然而，萨伦伯格并没有失去冷静，即使他意识到两台发动机都已关闭。

7. 在辅助动力装置（APU）的帮助下，飞机仍在上升，但明显减速。

8. 飞机在副驾驶员呼叫飞行速度和高度的同时，飞越了不到900英尺（约270米）的乔治·华盛顿大桥。

9. 驾驶舱内充斥着电脑反复提示的警告，"地势太低，太低，注意地形，拉升，拉升，拉升"。

10. 但自然，这还不是故事的结尾——机组人员必须从漂浮在河中间的飞机上疏散乘客。

Speaking

Pilot: Hello, I heard you're interested in aviation. Is that right?

Student: Yes, absolutely! I find aviation fascinating, especially when it comes to extraordinary incidents like the Miracle on the Hudson. It's incredible how Captain Sullenberger managed to land the plane safely on the river. I would love to learn more about it.

Pilot: It was indeed an incredible feat. The pilot's skill and quick thinking played a crucial role in saving all the passengers and crew. The incident happened when the plane struck a flock of geese and lost both engines. Captain Sullenberger had to make split-second decisions to ensure everyone's safety.

Student: That sounds extremely challenging. How did the pilot manage to glide the plane and land it on the Hudson River?

Pilot: Well, once the engines failed, Captain Sullenberger and his co-pilot, Jeffrey Skiles, realized they couldn't make it back to an airport. They made the decision to perform a controlled ditching in the river. With the engines out, they had to rely on the plane's momentum and glide it down towards the river's surface.

Student: That must have been nerve-wracking for everyone involved. How did they ensure the safety of the passengers during the landing?

Pilot: During the descent, Captain Sullenberger informed the passengers to brace for impact. The cabin crew quickly shouted commands to the passengers, instructing them to brace and stay down. The pilots focused on the landing, maneuvering the plane to avoid obstacles and aiming for a safe touchdown on the water.

Student: It's amazing how they managed to keep everyone calm and coordinated during such a critical moment. What happened after the landing?

Pilot: Once the plane came to a stop on the river's surface, the cabin crew initiated the evacuation process. Passengers were directed to exit onto the wings or onto rafts. The rescue operation was swift, with boats arriving promptly to assist. Although there were some challenges, such as water filling the plane through an open door, all passengers and crew were successfully evacuated.

Student: That's remarkable. I can only imagine the level of teamwork and coordination required to ensure everyone's safety. What was the aftermath of the incident?

Pilot: The successful landing on the Hudson River was recognized as one of the most successful ditching maneuvers in aviation history. Captain Sullenberger and the crew received numerous awards for their outstanding achievements. The event also gained significant media attention and was even adapted into a movie called "Sully: Miracle on the Hudson".

Student: It's truly an inspiring story. I'm amazed by the pilot's skill and the teamwork displayed during such a critical situation. It makes me appreciate the importance of pilot training and preparedness.

Pilot: Absolutely. This incident highlights the critical role of pilot training, decision-making, and teamwork in emergency situations. It reminds us of the importance of constant vigilance and preparedness in the aviation industry.

Student: Thank you for sharing these insights with me. It's been enlightening to learn about the Miracle on the Hudson from your perspective as a pilot.

Pilot: You're welcome! I'm glad I could share my knowledge with you. If you have any more questions about aviation or any other topic, feel free to ask.

Writing

Directions: *Write a composition entitled Only Stricter Traffic Laws Can Prevent Accidents. You should write at least 120 words according to the outline given below in Chinese:*

1. 每天全世界都有成千上万的人死于交通事故；
2. 分析产生交通事故的原因；
3. 应该制定更完善的交通法规遏止交通事故的发生。

Expansion Exercise for CET-BAND 4—News Report

At around 9:30 this morning, a trailer attached to a lorry turned over at the crossing of High Street in Milton. Hundreds of frozen turkeys were spilled all over the road. It is reported that nobody was hurt in the incident, but police said it may affect traffic and Christmas dinners. With just one week to go before Christmas, there are worries that local supermarket supplies of this holiday favorite may be affected. A police spokeswoman said that officers were currently in attendance at the scene. She stated that the driver of the lorry had been arrested on suspicion of dangerous driving. The crossing on High Street is a well-known accident black spot. This year alone, there have been seven traffic accidents at this location. Thankfully, none of these accidents have resulted in serious injury.

1. What does the news report say about the accident at the crossing of High Street in Milton? （　）

    A. It may have been due to the lorry driver's drunk driving.

    B. It may affect the local supply of turkeys for Christmas.

    C. It interrupted traffic for several hours running.

    D. It was caused by a lorry running into a trailer.

2. What do we learn about the crossing on High Street? （　）

    A. It has been the scene of several fatal accidents recently.

    B. It is the spot that causes the local police a lot of worry.

    C. It has witnessed several traffic accidents this year.

    D. It is a location frequented by local traffic police.

Unit 7 Aviation Technology

Preview

Artificial intelligence (AI) and its close relation to machine learning (ML) have reached unprecedented heights in the aviation industry. Both commercial airlines and military aviation have warmly embraced AI, utilizing its capabilities to streamline routes, reduce harmful emissions, enhance the customer experience, and optimize mission outcomes. However, along with its potential benefits, AI also presents a series of questions, technical challenges, and mixed sentiments.

Listening

Directions: *Fill in the blanks with the words that you hear. Find them in the table below and write down the corresponding letters.*

While AI has proven its __1__ in developing algorithms for personalized recommendations and daily news, using it for __2__ like flight path mapping, autonomous flying, or enabling __3__ UAVs raises trust issues, particularly among government entities and __4__.

One significant __5__ faced by the FAA and EASA in their discussions on AI is the absence of a standardized __6__ for this evolving technology. How can we define something that is constantly __7__? AI surpasses the __8__ of typical algorithms or __9__ used in everyday life, as it empowers machines to learn from experience and adapt their responses based on newly acquired __10__.

| A. evolving | B. value | C. challenge | D. complexity | E. definition |
| F. programs | G. swarming | H. tasks | I. data | J. consumers |

Reading

AI in Aviation: The Synergy of Artificial Intelligence and Flight Operations

Regulatory Interest in AI:

Recognizing the significance of AI, both the **Federal Aviation Administration** and the **European Union Aviation Safety Agency (EASA)** have taken a keen interest in the field. In February 2020, EASA published a report that **delved** into the trustworthiness of AI and **advocated** for a human-centric approach to AI programs in aviation. **Renowned** aviation manufacturers, such as Boeing and Airbus, are actively involved in AI research and development through individual efforts and international collaborations. Moreover, the **Society of Aerospace/Automotive Engineers (SAE)**, the world's aerospace safety organization, is actively publishing aviation standards and training materials based on AI, with AFuzion Inc. serving as the primary training resource for all SAE worldwide programs. Nonetheless, numerous unanswered questions, especially **pertaining** to safety, remain as AI continues to establish its place in our safety-critical world.

Defining AI:

One significant challenge faced by the FAA and EASA in their discussions on AI is the absence of a standardized definition for this evolving technology. How can we define something that is constantly evolving? AI **surpasses** the complexity of typical **algorithms** or programs used in everyday life, as it empowers machines to learn from experience and adapt their responses based on newly acquired data. Traditional aviation software **adheres** to **certification guidelines**, such as DO-178C for avionics software and DO-254 for avionics hardware, ensuring **deterministic** behavior. However, AI introduces an element of **variability**, as the same inputs can **yield** different outcomes over time as the software "learns". Ensuring **mandatory** certification **determinism** for evolving AI programs poses a significant challenge.

Safety Concerns and AI:

While AI has proven its value in developing algorithms for personalized recommendations and daily news, using it for tasks like flight path mapping, autonomous flying, or enabling **swarming UAV**s raises trust issues, particularly among government entities and consumers. EASA defines AI broadly as "any technology that appears to emulate the performance of a human," which highlights the human-like aspect of AI and raises questions about safety. While humans **are prone to** errors, it does not necessarily mean that AI will exhibit similar errors or **violate** safety **protocols**. Engineers have developed solutions for deterministic AI learning and real-time monitoring, although challenges remain in effectively communicating how AI operates to stakeholders, including passengers, pilots, and regulators. EASA and certification authorities are actively working towards **addressing** these challenges, **stimulating** international discussions, and developing **initiatives** to **tackle** the complex safety and **cybersecurity** aspects of AI-assisted aviation.

Investment and Future Prospects:

EASA and the aviation industry as a whole are increasing their investment in AI research and technology. They aim to encourage other countries and entities to **follow suit** and **incorporate** AI into their aviation industries. Notably, AI is already making its way into flight planning, **simulation**, and training applications, **paving the way** for its gradual **integration** into the **cockpit**. AFuzion predicts

that aviation AI will follow a **trajectory** similar to automobiles, with meaningful cockpit AI solutions expected to emerge in the 2030s.

Case Study: Alaska Airlines and AI:

Alaska Airlines serves as an excellent example of an airline **leveraging** AI to **streamline** and enhance aircraft performance. Utilizing the slowdown caused by the pandemic, the airline conducted a six-month **trial** period to test an AI-driven program called "Flyways." This program considered the original route, current weather conditions, aircraft weight, and other factors to determine the most efficient flight path. By continuously collecting and analyzing data on mileage and fuel usage, Flyways adapted and adjusted its efforts in real-time, resulting in an average time reduction of five minutes per flight. This seemingly small improvement translated into **immense** fuel savings, with Alaska Airlines saving a remarkable 480 thousand **gallons** of jet fuel during the trial period. This achievement **aligns** with the company's **commitment** to becoming **carbon-neutral** by 2040.

Conclusion:

The **utilization** of AI in aviation holds immense potential, with benefits **encompassing** efficiency, **sustainability**, and enhanced safety. While regulatory bodies, manufacturers, and industry experts are actively addressing safety concerns, **ongoing** collaboration and research are vital. The integration of AI into the aviation industry has the power to **revolutionize** air travel, making it more efficient and **sustainable**. As we continue to explore and embrace AI, we **embark** on a **transformative** journey towards a future where technology and aviation seamlessly **converge**.

New Words and Expressions

synergy *n.* /ˈsinədʒi/   the extra energy, power, success, etc. that is achieved by two or more people or companies working together, instead of on their own

协同增效作用

▲regulatory *adj.* /ˈreɡjələtəri/	having the power to control an area of business or industry and make sure that it is operating fairly （对工商业）具有监管权的，监管的
delve *v.* /delv/	search for sth. inside a bag, container, etc. （在手提包、容器等）翻找
▲advocate *v.* /ˈædvəkeɪt/	to support sth. publicly 拥护，支持，提倡
renowned *adj.* /rɪˈnaʊnd/	famous and respected 有名的，闻名的，受尊敬的
pertain *v.* /pəˈteɪn/	to exist or to apply in a particular situation or at a particular time 存在，适用
surpass *v.* /səˈpɑːs/	to do or be better than sb./sth. 超过，胜过，优于
algorithm *n.* /ˈælɡərɪðəm/	a set of rules that must be followed when solving a particular problem 算法，计算程序
adhere *v.* /ədˈhɪə(r)/	~ (to sth.) to stick firmly to sth. 黏附，附着
deterministic *adj.* /dɪˌtɜːrmɪˈnɪstɪk/	确定性的，命运注定论的
variability *n.* /ˌveəriəˈbɪləti/	the fact of sth. being likely to vary 可变性，易变性，反复不定
yield *v.* /jiːld/	to produce or provide sth., for example a profit, result or crop 出产，产生，提供
mandatory *adj.* /mænˈdeɪtəri/	required by law 强制的，法定的，义务的
determinism *n.* /dɪˈtɜːmɪnɪzəm/	the belief that people are not free to choose what they are like or how they behave, because these things are decided by their surroundings and other things over which they have no control 决定论（排除自由意志，认为个性或行为均由环

境和自己不能控制的因素所决定）

swarm *v.* /swɔːm/	to move around in a large group 成群地来回移动
emulate *v.* /ˈemjuleɪt/	to work in the same way as something else and perform the same tasks 仿真，模仿
prone *adj.* /prəʊn/	(~ to sth./to do sth.) likely to suffer from sth. or to do sth. bad 易于遭受，有做（坏事）的倾向
▲violate *v.* /ˈvaɪəleɪt/	to go against or refuse to obey a law, an agreement, etc. 违反，违背（法律、协议等）
protocol *n.* /ˈprəʊtəkɒl/	the first or original version of an agreement, especially a treaty between countries, etc.; an extra part added to an agreement or treaty 条约草案，议定书，（协议或条约的）附件
address *v.* /əˈdres/	to think about a problem or a situation and decide how you are going to deal with it 设法解决，处理，对付
▲**stimulate** *v.* /ˈstɪmjʊˌleɪt/	to encourage sth. to begin or develop further 鼓励，刺激
initiative *n.* /ɪˈnɪʃətɪv/	a new plan for dealing with a particular problem or for achieving a particular purpose 倡议，新方案
tackle *v.* /ˈtækəl/	to deal with a difficult problem or task in a very determined or efficient way. 处理
incorporate *v.* /ɪnˈkɔːpəreɪt/	to include sth. so that it forms a part of sth. 将……包括在内，包含，吸收，使并入
simulation *n.* /ˌsɪmjuˈleɪʃn/	a situation in which a particular set of conditions is created artificially in order to study or experience sth. that could exist in reality

	模拟,仿真
▲**integration** *n.* /ˌɪntɪˈɡreɪʃn/	the act or process of combining two or more things so that they work together (= of integrating them)
	结合,整合,一体化
cockpit *n.* /ˈkɒkpɪt/	the area in a plane, boat or racing car where the pilot or driver sits
	(飞机、船或赛车的)驾驶舱,驾驶座
trajectory *n.* /trəˈdʒektəri/	the curved path of sth. that has been fired, hit or thrown into the air
	(射体在空中的)轨道,弹道,轨迹
leverage *v.* /ˈliːvərɪdʒ/	use (something) to maximum advantage
	最大限度地利用,最优化使用
streamline *v.* /ˈstriːmlaɪn/	to give sth. a smooth even shape so that it can move quickly and easily through air or water
	使成流线型
trial *n.* /ˈtraɪəl/	the process of testing the ability, quality or performance of sb./sth.
	(对能力、质量、性能等的)试验,试用
▲**immense** *adj.* /ɪˈmens/	extremely large or great
	极大的,巨大的
gallon *n.* /ˈɡælən/	a unit for measuring liquid; in the UK, Canada and other countries it is equal to about 4.5 litres; in the US it is equal to about 3.8 litres; there are four quarts in a gallon
	加仑(液量单位,在英国、加拿大等国约等于4.5升,在美国约等于3.8升,一加仑为4夸脱)
align *v.* /əˈlaɪn/	(~ ... or ~ with/to ...) change sth. so it is similar to sth. else; join some group or cooperate with sb. else
	调整,使一致,与……结盟
▲**commitment** *n.* /kəˈmɪtmənt/	a promise to do sth. or to behave in a particular way; a promise to support sb./sth.;

	the fact of committing yourself
	承诺，许诺，承担，保证
carbon-neutral *adj.*	a type of lifestyle that does not cause an increase in the overall amount of carbon dioxide in the atmosphere
	碳中和的
utilization *n.* /ˌjuːtələˈzeɪʃn/	use
	利用，使用
encompass *v.* /ɪnˈkʌmpəs/	to include a large number or a range of things
	包含，包括，涉及（大量事物）
▲sustainability *n.* /səˌsteɪnəˈbɪləti/	the property of a long-term supply
	可持续性
ongoing *adj.* /ˈɒŋɡəʊɪŋ/	continuing to exist or develop
	持续存在的，仍在进行的，不断发展的
revolutionize *v.* /ˌrevəˈluːʃənaɪz/	to completely change the way that sth. is done
	彻底改变，完全变革
sustainable *adj.* /səˈsteɪnəbl/	involving the use of natural products and energy in a way that does not harm the environment; that can continue or be continued for a long time
	不破坏生态平衡的，合理利用的，可持续的
embark *v.* /ɪmˈbɑːk/	to get onto a ship; to put sth. onto a ship
	上船，装船
transformative *adj.* /trænsˈfɔːrmətɪv/	有改革能力的，起改造作用的
converge *v.* /kənˈvɜːdʒ/	to move towards a place from different directions and meet
	汇集，聚集，集中
Federal Aviation Administration (FAA)	联邦航空局
European Union Aviation Safety Agency (EASA)	欧洲航空安全局
Society of Aerospace/Automotive Engineers (SAE)	航空航天/汽车工程师学会
certification guideline	认证指南
Unmanned Aerial Vehicle (UAV)	无人机

follow suit	效仿，跟随他人的做法或行动
cyber security	网络安全
pave the way	为……作准备，为……铺平道路

Genre Analysis

Genre:

The genre of this passage is informational or expository. It provides information about integrating AI into aviation and discusses topics like regulations, safety concerns, investment trends, and case studies.

Communicative moves:

Move 1 Background and Regulatory Interest:

Providing background information on the involvement of regulatory bodies in AI research and development in aviation.

Move 2 Defining AI:

Exploring the challenge of defining AI in the context of aviation and highlighting its complexities compared to traditional algorithms.

Move 3 Safety Concerns:

Discussing safety concerns associated with AI in aviation and emphasizing the need for addressing trust issues, effective communication of AI operation, and adherence to safety protocols.

Move 4 Investment and Future Prospects:

Highlighting the increasing investment in AI research and technology in the aviation industry and predicting the gradual integration of AI into the cockpit.

Move 5 Case Study:

Alaska Airlines and AI: Presenting a specific case study of Alaska Airlines using AI to optimize aircraft performance and showcasing the practical benefits and fuel savings achieved through AI implementation.

Move 6 Conclusion:

Summarizing the potential benefits of AI in aviation and emphasizing the need for ongoing collaboration, research, and the transformative impact AI can have on air travel.

Communicative Purpose:

The purpose of the text is to inform readers about the current state, challenges, and potential of integrating AI into flight operations in the aviation industry. It aims to highlight the regulatory interest, safety concerns, investment trends, and real-world case studies to demonstrate the benefits and transformative power of AI in aviation. The text seeks to foster understanding and stimulate further research, collaboration, and adoption of AI technologies in the aviation industry.

Culture Tips

1. Stay up-to-date with regulatory guidelines: In the aviation industry, it's important to stay informed about the evolving regulations and guidelines set by regulatory bodies. This shows a commitment to safety and compliance.
2. Embrace collaboration and research: The passage highlights the importance of collaboration and ongoing research in the field of AI in aviation. Embrace a culture of collaboration, knowledge-sharing, and continuous learning to stay at the forefront of technological advancements.
3. Foster effective communication: Communication plays a vital role in addressing safety concerns and building trust in AI technologies. Emphasize clear and transparent communication about AI operations to stakeholders, including passengers, pilots, and regulators. This helps to alleviate concerns and ensure a shared understanding of AI implementation.
4. Emphasize safety and reliability: Safety is paramount in the aviation industry. Prioritize safety and reliability when integrating AI into flight operations. This includes rigorous testing, validation, and adherence to safety protocols to ensure AI systems make safe decisions in real-time.
5. Promote efficiency and sustainability: Efficiency and sustainability are key benefits of AI in aviation. Foster a culture that values efficiency, seeks optimization opportunities, and aligns with sustainability goals. Encourage initiatives that reduce fuel consumption and carbon emissions, such as optimizing flight paths and reducing operational inefficiencies.
6. Embrace technological advancements: Embrace the potential of AI and other technological advancements in the aviation industry. Stay open to exploring new possibilities, investing in research and development, and continuously adapting to emerging technologies that can enhance flight operations.
7. By embracing these culture tips, individuals and organizations can create an environment that encourages safe, efficient, and sustainable integration of AI in aviation while fostering collaboration, communication, and innovation.

Reading Exercises

I. Vocabulary Matching

Directions: *Fill in the blank with the corresponding letter of the correct word from the box for each sentence.*

1. Recognizing the _____, regulatory bodies have shown interest in AI programs in aviation.
2. The definition of AI in aviation and the challenges of ensuring certification _____ remain unanswered.
3. Safety concerns related to the use of AI in aviation and its implications for _____ are raised.
4. The passage presents a predicted trajectory for aviation AI _____ and increased investment.
5. The case study highlights how Alaska Airlines used an AI-driven program to enhance _____ performance.
6. The _____ of AI in aviation poses complex challenges for regulation and certification.
7. Effective _____ of AI operations to stakeholders is crucial for ensuring safety and understanding.
8. The _____ of AI in flight planning and simulation offers potential for efficiency and training.
9. _____ questions about AI's definition and certification determinism in aviation remain unanswered.
10. The passage highlights the increased _____ in AI research and technology by EASA and the industry.
11. The case study demonstrates the _____ of AI in determining efficient flight paths.
12. The passage mentions the publication of a _____ by EASA and collaboration with aviation manufacturers.
13. The passage discusses the complexity of defining AI in aviation and ensuring _____ in evolving programs.
14. The _____ of AI in aviation presents opportunities and challenges for the industry and regulators.
15. The passage emphasizes the need for ongoing collaboration, research, and addressing safety _____.

A. safety B. significance C. determinism D. development E. aircraft
F. integration G. concerns H. communication I. regulatory J. trustworthiness
K. Numerous L. investment M. efficiency O. report P. certification
Q. impact

II. Answer the Following Questions

1. What are some benefits of utilizing AI in aviation?

2. What challenges are faced in defining AI in the aviation industry?

3. What safety concerns arise when implementing AI in aviation?

4. How are regulatory bodies and industry players approaching AI in aviation?

5. Can you provide an example of an airline utilizing AI in aviation operations?

III. Multiple Choice

1. Which of the following is a benefit of utilizing AI in aviation? ( )

    A. Increased ticket prices.

    B. Longer flight duration.

    C. Streamlining routes and reducing emissions.

    D. Decreased customer satisfaction.

2. What is one challenge in defining AI in the aviation industry? ( )

    A. Lack of interest from regulatory bodies.

    B. Standardization of AI definitions.

    C. Certification determinism for traditional algorithms.

    D. Inability of AI to adapt to new data.

3. What safety concerns arise when implementing AI in aviation? ( )

    A. Potential for AI to violate safety protocols.

    B. AI's inability to emulate human performance.

    C. Lack of interest from stakeholders.

    D. Certainty of AI learning outcomes.

4. How are regulatory bodies and industry players approaching AI in aviation? ( )

    A. Ignoring the potential of AI.

    B. Investing heavily in AI research and development.

    C. Avoiding collaboration with manufacturers.

    D. Limiting discussions on AI safety and cybersecurity.

5. Can you provide an example of an airline utilizing AI in aviation operations? ( )

    A. Delta Airlines implementing AI for flight bookings.

B. Southwest Airlines introducing AI for inflight entertainment.

C. Alaska Airlines testing an AI-driven program called "Flyways".

D. United Airlines using AI for cabin crew scheduling.

Translation

Directions: *Translate the following sentences into Chinese.*

1. Alaska Airlines used AI to optimize flight paths and achieve fuel savings and carbon neutrality.

2. The FAA and EASA are regulatory bodies that show interest in the development of AI in aviation and are working on addressing its trustworthiness and human-centric approach.

3. Defining AI in the aviation context is challenging due to its evolving nature, as it allows machines to learn from experience and adapt responses based on new data.

4. Alaska Airlines optimized flight paths through its AI-driven program called Flyways, which tested different routes, collected data, and adjusted efforts to create real-time efficient flight paths.

5. The future outlook for AI in the aviation industry is promising, as the industry continues to invest in research and technology, gradually introducing AI into the cockpit, following a similar path as the automotive industry.

6. The FAA and EASA are exploring AI's safety concerns, including its definition, safety considerations, and clear communication of AI operations in aviation.

7. Meaningful cockpit AI solutions can be anticipated in the 2030s, according to AFuzion, an aerospace safety organization.

Speaking

Student: I've been reading about AI in aviation. Why are regulatory bodies interested in it?

AI expert: Regulatory bodies ensure safety in aviation. With AI's potential impact, they want to address safety concerns and establish guidelines for its integration.

Student: What are the safety concerns with AI in aviation?

AI expert: Trust and reliability of AI systems are key concerns. We need to ensure they make safe decisions in real-time and communicate their operations effectively.

Student: Can you explain the case study about Alaska Airlines using AI?

AI expert: Alaska Airlines used an AI program called "Flyways" to find the most efficient flight paths. It reduced flight times by five minutes, saving 480,000 gallons of fuel during the trial period.

Student: What's the future for AI in aviation?

AI expert: The industry is investing more in AI research. Meaningful AI solutions in the cockpit could emerge in the 2030s, revolutionizing air travel for efficiency, sustainability, and safety.

Student: Thanks for the information! It's exciting to see the potential of AI in aviation.

AI expert: You're welcome! If you have more questions, feel free to ask. AI in aviation is a fascinating field with lots of possibilities.

Writing

Directions: *For this part, you are allowed 30 minutes to write an essay about the importance of developing a healthy lifestyle among college students. You should write at least 120 words but no more than 180 words.*

Expansion Exercise for CET-BAND 4—News Report

Despite smartphones and social media, young people today are as socially competent as those from the previous generation. At least this is what a new study suggests. For the study, researchers compared teacher and parent evaluations of American children who started kindergarten in 1998 with those who began school in 2010. The former group entered kindergarten when mobile phones were luxuries. The latter group started school when mobile devices were widespread. Results showed both groups of children were rated similarly on important social skills. These included the ability to form and maintain friendships and get along with people who were different. They were also rated similarly on self-control, such as the ability to regulate their temper.

In virtually every comparison made, ratings of social skills either remained constant or improved for the children born later. There was one exception: Social skills were slightly lower for children who accessed online games and social networking sites many times a day.

Adults are worried when technological change starts to undermine traditional relationships, particularly the parent-child relationship. The introduction of telephones, automobiles and radio all led to moral panic among adults of the time because the technology allowed children to enjoy more freedom. Fears over screen-based technology represent the most recent panic in response to technological change. But overall, the study found little evidence that time spent on screens was hurting social skills for most children.

(2022.9-P2)

1. What does the new study suggest about young people today and those from the previous generation? (　)

   A. Both groups spend a lot of time on mobile devices.

   B. Both groups attach importance to social connections.

   C. They are equally competent in using new technology.

   D. They are similar in terms of social skills.

2. What did the study find about children who access social networking sites many times a day?（　）

   A. Their social skills were negatively affected.

   B. Their school performance was slightly lower.

   C. Their emotions were much harder to regulate.

   D. Their relations with peers were badly strained.

3. What is adults' worry about technological change?（　）

   A. It may pose a threat to their children's safety.

   B. It may affect society's traditional values.

   C. It may hurt their relations with children.

   D. It may change their children's ethical values.

Unit 8　Aviation Anecdote

Preview

　　The passage of this unit is about Justin Mutawassim, a Delta Air Lines pilot whose tweet became popular. It shares his journey of becoming a pilot, starting from his childhood dream. Despite obstacles, he remained determined and worked at Delta as a ramp agent. With the guidance of a mentor, he attended flight school, became a flight instructor, and eventually got a pilot job at Breeze Airways. When Delta removed the college degree requirement for pilots, he applied and got accepted. His celebratory tweet gained attention, highlighting the importance of mentorship and creating an inspiring workplace. Delta Air Lines is committed to diversity and removing barriers for talented pilots.

Listening

Directions: *Fill in the blanks with the words that you hear. Find them in the table below and write down the corresponding letters.*

This is a story about Delta Air Lines, dreams, and mentorship. More than 20 years ago, a seven-year-old boy was __1__ to meet the pilot and see the cockpit in Delta Air Lines. From that moment, he later said, he knew __2__ what he wanted to be when he grew up. "I remember sitting there and being __3__ by all the buttons," said Justin Mutawassim, who is now 26 years old.

And although a teacher derailed his dream temporarily by telling him incorrectly that airline pilots had to have uncorrected 20/20 vision, Mutawassim, who wears glasses and had no easy way to check that __4__ at the time, recovered. After high school, he took a job as close to flying as he could get without a degree or pilot's __5__: as a runway agent for Delta. A few years later, on an airport employee bus, he struck up a __6__ with Ivor Martin, a Virgin America pilot, who said that while he loved working as a runway agent, he really wanted to be a pilot. So Martin helped Mutavasim lay out the practical steps he would have to take to __7__ his dream.

In 2016, Mutawassim left his Delta job for flight school, taking out loans and __8__ the program in 11 months. He worked as a flight instructor, and then as a pilot for a regional airline, as he accumulated flight hours. Last summer, he was hired as a pilot at startup Breeze Airways. But earlier this year, Delta __9__ it was dropping the requirement that pilots have college degrees, and Mutawassim applied. He went through training and finished his final __10__ late last month.

Finally, this story tells us, as long as we have a dream in our heart, we should boldly pursue it!

| A. qualifications | B. fascinated | C. exactly | D. announced | E. license |
| F. invited | G. information | H. achieve | I. conversation | J. finishing |

Reading

A Delta Air Lines Pilot's Inspiring Tweet Went Viral

This is a story about Delta Air Lines, dreams, and **mentorship**. It's also about social media and inspiring the **employees** you'll want to **recruit** down the road. More than 20 years ago, a young boy—5, 6, then 7 years old—would **routinely** travel back and forth between New York and Texas to visit his family. He was invited to meet the pilots and visit the **cockpit** on one such flight on Delta Air Lines. From that moment, he later said, he knew exactly what he wanted to be when he grew up.

"I remember sitting there and being **fascinated** by all the **buttons**," said Justin Mutawassim, who is now 26 years old. "From there, I just caught the **bug**." Throughout his youth, Mutawassim **was known for** his single-minded focus. ("Remember when I would talk **nonstop** about flying planes for a living in school?" Mutawassim wrote to friends on Twitter four years ago.) And although a teacher **derailed** his dream **temporarily** by telling him incorrectly that airline pilots had to have uncorrected 20/20 vision, Mutawassim, who wears glasses and had no easy way to check that information at the time, **recovered**.

After high school, he went to community college **briefly** in Dallas in 2014, but then left to get a job as close to flying as he could with no degree or pilot's license: working as a **ramp agent** for Delta. "I just absolutely fell in love with the technical **aspect** of **aviation**," Mutawassim said. "It was physically the hardest job I've ever done. Manual labor is no joke."

Then, he got the first of a few big breaks. On an airport employee bus, he struck up a conversation with Ivor Martin, who was then a pilot for Virgin America, and mentioned that while he liked being a ramp agent, he really wanted to be a pilot. The single exchange developed into a mentorship relationship, as Martin helped Mutawassim **plot out** the practical steps he'd have to take to make his dream a reality.

In 2016, Mutawassim left his Delta job for flight school, taking out **loans** and finishing the program in 11 months. He worked as a flight instructor, and then as a pilot for a **regional** airline, as he **accumulated** flight hours. Last summer, he was hired as a pilot at startup Breeze Airways.

But earlier this year, Delta announced it was dropping the requirement that pilots have college degrees, and Mutawassim applied. He went through training and finished his final **qualifications** late last month, and then posted a **tweet** about his accomplishment:

"This one has been 6 long years in the making. How it started. How it's going. pic. twitter.com/wXqN8lITE3–Never Enough Legroom (@j_muta) September 30, 2022"

Who wouldn't be **romantic** about careers in air travel after a post like that? Sure enough, it blew up: About 164,000 likes as of this writing, along with lots of other people sharing their photos and stories of what they were doing for work years ago, compared to the better-loved things they're doing now.

"Uhhhhhh. Hi everyone," Mutawassim wrote in a later tweet. "Was only expecting my 10 friends who still use this app to see this. Welcome. If you're interested. Please consider donating to the Organization of Black Aerospace Professionals."

Mutawassim is certainly getting some attention for all of this, which is great. I first heard about all of this in Sydney Page's article in The Washington Post, and besides Mutawassim's tweet, his post on LinkedIn went viral (well, "**viral**-for-LinkedIn"), as well.

But I think the real lesson if you're a business owner comes in two parts.

First, it's to recognize people like Martin, the pilot for Virgin America (he's now with Alaska Airlines) who took the time to mentor Mutawassim, and even the Delta pilots back in 2000 or so who **originally** inspired his dream.

And second, it's to ask yourself **a couple of** questions:

"Would you like it if people took to social media, posting about how working for you was a lifelong dream, and the posts then went viral? If so, what do you have to do now, in order to make that happen down the road?"

I asked Delta Air Lines for comment on Mutawassim's story. Here's the **statement** they sent me:

"Delta's **commitment** to developing a diverse workforce that is **reflective** of the communities we serve **domestically** and around the world is unwavering. This includes removing barriers and **broadening** recruiting **funnels** to help create a **diverse pipeline** for qualified and talented pilots to join us. One way we have done so is through our Propel Career Pathway Program which **launched** in 2018. Propel offers **multiple pathways** for current Delta employees and students at our 15 partner universities to reach the **flight deck** of a Delta jet. Within Propel's Company Pathway, we currently have more than 100 employees at various stages of the program, including those who have experience as **flight attendants**, ramp workers, **gate agent**s, mechanics, **flight dispatchers** and other administrative or support roles at Delta."

New Words and Expressions

mentorship *n.* /ˈmentəʃɪp/   a system where seniors guide new people to do a certain work
导师，师徒制，师友计划

employee *n.* /ɪmˈplɔɪiː/   a person who is paid to work for sb.
受雇者，雇工，雇员

▲recruit *v.* /rɪˈkruːt/   to find new people to join a company, an organization, the armed forces, etc.
吸收（新成员），征募（新兵）

routinely *adv.* /ruːˈtiːnli/   according to an established practice
按常规，例行公事地，照例

▲cockpit *n.* /ˈkɒkˌpɪt/   the area in a plane, boat or racing car where the pilot or driver sits
（飞机、船或赛车的）驾驶舱，驾驶座

fascinate *v.* /ˈfæsɪneɪt/   to attract or interest sb. very much
深深吸引，迷住

button *n.* /ˈbʌt(ə)n/   a small part of a machine that you press to make it work
按钮

bug *n.* /bʌg/   an enthusiastic interest in sth. such as a sport or a hobby
热衷，着迷

nonstop *adj.* /nɒnˈstɒp/   without stop
不停的，不断的，（列车、飞机等）直达的

derail *v.* /diːˈreɪl/   to leave the track; to make a train do this; obstruct (a process) by diverting it from its intended course
脱轨，出轨，使脱离原进程，阻碍，阻挠

▲temporarily *adv.* /ˈtemprərəli/   lasting or intended to last or be used only for a short time; not permanent
暂时，临时

▲recover *v.* /rɪˈkʌvə(r)/   to get well again after being ill/sick, hurt, etc.

恢复健康，康复，痊愈

ramp *n.* /ræmp/ — a slope or set of steps that can be moved, used for loading a vehicle or getting on or off a plane
（装车或上下飞机的）活动梯，活动坡道

▲aspect *n.* /ˈæspekt/ — a particular part or feature of a situation, an idea, a problem, etc.; a way in which it may be considered
方面，层面

aviation *n.* /ˌeɪviˈeɪʃ(ə)n/ — the designing, building and flying of aircraft
航空制造业，航空

▲conversation *n.* /ˌkɒnvə(r)ˈseɪʃ(ə)n/ — an informal talk involving a small group of people or only two; the activity of talking in this way
交谈，谈话

loan *n.* /ləʊn/ — money that an organization such as a bank lends and sb. borrows
贷款，借款

regional *adj.* /ˈri:dʒ(ə)nəl/ — of or relating to a place
地区的，区域的，地方的

▲accumulate *v.* /əˈkju:mjʊleɪt/ — to gradually get more and more of sth. over a period of time
积累，积聚

qualification *n.* /ˌkwɒlɪfɪˈkeɪʃ(ə)n/ — an exam that you have passed or a course of study that you have successfully completed
（通过考试或学习课程取得的）资格，学历

tweet *n.* /twi:t/ — a message sent using the Twitter social networking service
运用推特社交网络发送的信息

romantic *adj.* /rəʊˈmæntɪk/ — connected or concerned with love or a sexual relationship
浪漫的，爱情的，情爱的

viral *adj.* /ˈvaɪrəl/ — like or caused by a virus; relating to or involving the rapid spread of information

	about a product or service by viral marketing techniques 病毒的，病毒性的，病毒引起的，利用病毒式营销手段传播的
originally *adv.* /əˈrɪdʒ(ə)nəli/	at the beginning of a particular period, process or activity 原来地，最初地
statement *n.* /ˈsteɪtmənt/	a formal or official account of facts or opinions 声明，陈述，报告
commitment *n.* /kəˈmɪtmənt/	a promise to do sth. or to behave in a particular way; a promise to support sb./sth. 承诺，许诺，允诺承担，保证
reflective *adj.* /rɪˈflektɪv/	thinking deeply about things 沉思的，深思的
domestically *adv.* /dəˈmestɪkli/	of or inside a particular country; not foreign or international 本国地
broaden *v.* /ˈbrɔːd(ə)n/	to become wider 变宽，变阔
funnel *n.* /ˈfʌn(ə)l/	a device that is wide at the top and narrow at the bottom, used for pouring liquids or powders into a small opening 漏斗
diverse *adj.* /daɪˈvɜːs/	very different from each other and of various kinds 不同的，相异的，多种多样的，形形色色的
pipeline *n.* /ˈpaɪplaɪn/	a series of pipes that are usually underground and are used for carrying oil, gas, etc. over long distances; a channel supplying goods or information 输送管道，供应货物、信息等的渠道
▲launch *v.* /lɔːntʃ/	to start an activity, especially an organized one 开始从事，发起，发动（尤指有组织的活动）

multiple *adj.* /ˈmʌltɪp(ə)l/	many in number; involving many different people or things
	数量多的，多种多样的
pathway *n.* /ˈpɑːθweɪ/	a way to sth.
	路径，通路，途径
current *n.* /ˈkʌrənt/	happening now; of the present time
	现时发生的，当前的，现在的
attendant *n.* /əˈtendənt/	a person whose job is to serve or help people in a public place
	服务员，侍者
be known for	因……而众所周知
ramp agent	机坪操作员
plot out	划分，分配，描绘，提出……纲要
a couple of	几个，一对，两个，两三个
flight deck	飞机驾驶舱
gate agent	机场登机口工作人员
flight dispatcher	飞行调度员

Genre Analysis

Genre:
The genre of this passage is a news article or feature story.

Communicative Moves:

Move 1 Introduction:
The article introduces the story of a Delta Air Lines pilot and his journey towards achieving his dream of becoming a pilot.

Move 2 Background Information:
The article provides background information about the pilot's childhood fascination with airplanes and his determination to pursue a career in aviation.

Move 3 Obstacles and Mentorship:
The article discusses the obstacles the pilot faced, such as a teacher's incorrect information about vision requirements, and how he overcame them with the help of a mentor who guided him towards his goal.

Move 4 Career Progression:
The article describes the pilot's career progression, from working as a ramp agent to

attending flight school, becoming a flight instructor, and eventually being hired as a pilot.

Move 5 Viral Tweet:

The article highlights the pilot's tweet about his accomplishment, which went viral on social media, garnering attention and support from the public.

Move 6 Lessons for Business Owners:

The article draws lessons for business owners, emphasizing the importance of mentorship and creating an environment where employees are inspired to share their positive experiences.

Move 7 Delta's Commitment to Diversity:

The article includes a statement from Delta Air Lines, highlighting the company's commitment to diversity and creating opportunities for aspiring pilots from diverse backgrounds.

Communicative Purpose:

The purpose of this article is to share an inspiring story of a pilot's journey and to emphasize the importance of mentorship, diversity, and creating a supportive work environment. It also aims to encourage business owners to consider the impact of their company culture and how it can lead to positive experiences and viral endorsements from employees. Additionally, the article showcases Delta Air Lines' commitment to diversity in pilot recruitment and highlights their Propel Career Pathway Program as an example of their efforts.

Culture Tips

1. Mentorship and guidance: The passage highlights the importance of mentorship in achieving one's goals. In many cultures, seeking guidance from experienced individuals and mentors is highly valued. Emphasize the significance of finding mentors who can provide support, advice, and direction in the student's journey.

2. Perseverance and determination: The pilot's story showcases the importance of perseverance and determination in overcoming obstacles and achieving dreams. Encourage the students to adopt a resilient mindset and to stay motivated even when faced with challenges.

3. Sharing knowledge and experiences: The pilot's willingness to share their story and offer guidance reflects a culture of sharing knowledge and experiences. Encourage the students to be open to learning from others and to share their own experiences with fellow enthusiasts or aspiring aviators.

4. Recognizing opportunities: The dialogue highlights how the pilot recognized an

opportunity to strike up a conversation with a pilot on an airport employee bus. Encourage the student to be observant and proactive in identifying opportunities that can help them progress in their aviation career.

5. Creating a supportive environment: The pilot's mention of having people around who believed in their potential emphasizes the importance of a supportive environment. Encourage the student to surround themselves with positive influences, whether it's friends, family, or like-minded individuals who can provide encouragement and support.

6. Embracing diversity: The passage mentions the pilot's affiliation with the Organization of Black Aerospace Professionals. This highlights the importance of diversity and inclusivity in the aviation industry. Encourage the student to embrace diversity, respect different perspectives, and contribute to creating an inclusive and welcoming aviation community.

7. Lifelong learning: The pilot's advice to never stop learning and improving skills underscores the value of continuous learning and professional development. Encourage the student to seek opportunities for learning, whether through formal education, workshops, seminars, or staying updated with industry advancements.

Reading Exercises

I. Vocabulary Matching

Directions: *Fill in the blank with the corresponding letter of the Chinese equivalent of the following English words.*

1. recruit ( ) 7. agent ( ) 13. career ( ) 19. multiple ( )
2. employee ( ) 8. aspect ( ) 14. recognize ( ) 20. dispatch ( )
3. cockpit ( ) 9. physically ( ) 15. comment ( ) 21. administrative ( )
4. fascinate ( ) 10. labor ( ) 16. inspire ( ) 22. be known for ( )
5. temporarily ( ) 11. accumulate ( ) 17. diverse ( ) 23. plot out ( )
6. community ( ) 12. accomplish ( ) 18. launch ( )

A. 代理人，经纪人　　G. 完成　　　　　　M. 方面，层面　　　S. 不同的，相异的
B. 身体上　　　　　　H. 议论，评论　　　N. 数量多的　　　　T. 激励，鼓舞
C. 劳动　　　　　　　I. 深深吸引，迷住　O. 派遣，调遣　　　U. 开始从事，发起
D. 积累，积聚　　　　J. 暂时，临时　　　P. 吸收，征募　　　V. 认识，认出
E. 驾驶舱　　　　　　K. 社区，社会　　　Q. 受雇者　　　　　W. 管理的，行政的
F. 划分，分配，描绘　L. 因……而众所周知　R. 事业

II. Answer the Following Questions

1. How did Justin Mutawassim become interested in becoming a pilot?
2. How did Justin Mutawassim pursue his dream of becoming a pilot?
3. How did Justin Mutawassim's tweet about becoming a Delta Air Lines pilot go viral?
4. What organization did Justin Mutawassim encourage donations to in his tweet?
5. What is Delta Air Lines' approach to developing a diverse pilot workforce?

III. Multiple Choice

1. Which job does Mutawassim want to do as a boy? (　)
    A. Maintenance.
    B. Cabin crew.
    C. Traffic controller.
    D. Pilot.
2. What was Mutavasim's first job? (　)
    A. A ramp agent for Delta.
    B. A maintenance person for Delta.
    C. An air traffic controller.
    D. A pilot for Delta.
3. Why did Mutavasim become a pilot in Delta? (　)
    A. Because Delta announced it was dropping its requirement for pilots to be licensed.
    B. Because Delta announced it was dropping its requirement for pilots to have flight experience.
    C. Because Delta announced it was dropping its requirement that pilots have a college degree.
    D. Because Delta announced that anyone can interview a pilot.
4. What can we learn about Propel Career Pathway Program? (　)
    A. It found many employment channels for graduates.
    B. It helps Delta to create a diverse pipeline for qualified and talented pilots.
    C. It creates a lot of business value.
    D. It diversifies the forms of community service.
5. Which of the following statement is not true? (　)
    A. Mutawassim was invited to meet the pilots and visit the cockpit when he was young.
    B. Mutawassim didn't get a college diploma.
    C. Mutawassim was eventually hired as a pilot by Breeze Airways.
    D. Mutawassim fell in love with the technical aspect of aviation.

Translation

Directions: *Translate the following sentences into Chinese.*

1. Within Propel's Company Pathway, we currently have more than 100 employees at various stages of the program, including those who have experience as flight attendants, ramp workers, gate agents, mechanics, flight dispatchers and other administrative or support roles at Delta.

   _____
   _____

2. On an airport employee bus, he struck up a conversation with Ivor Martin, who was then a pilot for Virgin America, and mentioned that while he liked being a ramp agent, he really wanted to be a pilot.

   _____
   _____

3. The single exchange developed into a mentorship relationship, as Martin helped Mutawassim plot out the practical steps he'd have to take to make his dream a reality.

   _____
   _____

4. Delta's commitment to developing a diverse workforce that is reflective of the communities we serve domestically and around the world is unwavering.

   _____
   _____

Speaking

Pilot: Good morning! Are you interested in aviation?

Student: Yes, I am! I've always been fascinated by airplanes and dreamt of becoming a pilot one day.

Pilot: That's wonderful! I can relate to your passion. You know, when I was your age, I had the same dream. I remember being your age and flying on Delta Air Lines. I was invited to the cockpit, and that experience ignited my desire to become a pilot.

Student: Wow, that must have been amazing! So, how did you make your dream come true?

Pilot: Well, it wasn't always smooth sailing. I faced some challenges along the way. One of my teachers wrongly told me that pilots had to have uncorrected 20/20 vision, and I wear glasses. But I didn't let that discourage me. I worked hard, sought guidance, and eventually found a mentor who helped me plot out the steps I needed to take.

Student: That's inspiring! How did you start your journey towards becoming a pilot?

Pilot: After high school, I started working as a ramp agent for Delta Air Lines. It was physically demanding, but I loved being close to aviation. One day, while on an airport employee bus, I struck up a conversation with a pilot from another airline. I shared my dream with him, and he became my mentor, guiding me through the practical steps I needed to take.

Student: That's amazing that you found a mentor who believed in you! What came next?

Pilot: I saved up and enrolled in flight school. It was a challenging but rewarding experience. After completing the program, I worked as a flight instructor and gained valuable flight hours. Eventually, I got hired as a pilot at a regional airline and recently, I fulfilled my dream of becoming a pilot for Delta Air Lines.

Student: That's incredible! Your journey is truly inspiring. I hope I can follow in your footsteps. Any advice for someone like me who wants to pursue a career in aviation?

Pilot: Absolutely! First and foremost, never give up on your dream. It may seem challenging at times, but with dedication and perseverance, you can achieve it. Surround yourself with people who believe in your potential, and don't hesitate to seek guidance from mentors or professionals in the field. And most importantly, never stop learning and improving your skills.

Student: Thank you so much for sharing your story and advice. I feel motivated and encouraged to pursue my passion for aviation. I'll work hard and stay determined to make my dream come true, just like you did.

Pilot: You're welcome! I have no doubt that with your passion and dedication, you'll soar to great heights. Remember, the sky's the limit! If you ever have any questions or need guidance along the way, feel free to reach out to me. Good luck on your journey, future aviator!

Writing

Directions: *Suppose the university newspaper is inviting submissions from the students for its coming edition on a campus event that has impressed them most. You are now to write an* *essay for submission. You will have 30 minutes to write the essay. You should write at least 120 words but no more than 180 words.*

Expansion Exercise for CET-BAND 4—Reading

Directions: *In this section, you are going to read a passage with ten statements attached to it. Each statement contains information given in one of the paragraphs. Identify the paragraph from which the information is derived. You may choose a paragraph more than once. Each paragraph is marked with a letter*

How to Not Be Boring

A. Humans are creatures of habit. We love to establish a routine and stick with it. Then we often put ourselves on auto-pilot. Routines can be incredibly useful in helping you get things done. However, too much of a routine can also make you incredibly boring. Nevertheless, many people live lives that are boringly predictable, or live a life where everything is outlined or planned.

B. To tell the truth, interesting people are more popular among their friends. If you don't arouse someone's curiosity or brighten someone's day you probably come across as being a little bit dull. But that doesn't mean your life has ended and you can't do anything to change it. If you find yourself searching for something to say beyond small talk, try these tactics to find more interesting approaches to conversation.

C. Recently, I was at a gathering of colleagues when someone turned to me and asked "So, what's new with you?" Ordinarily, I think I'm a good conversationalist. After all it's literally my job to talk to people and tell their stories or share their advice. And that's not exactly an unexpected question. Still, the only "new-to-me topics that came to mind were my daughter's basketball tournament（锦标赛）and my feelings about that morning's political headlines—neither amusing nor appropriate topics at that moment.

D. Oh, no, I thought. Have I become boring? But sharing our experiences in an authentic way to connect with other people is what makes us interesting, says Associate Professor Michael Pirson. The hesitation I felt in not sharing the ordinary things that were happening in my life, and the wild mental search for something more interesting, may have backfired and made me seem less interesting.

E. "If someone is making up some conversation that might be interesting, it's probably not going to land well," says Pirson, whose expertise includes trust and well-being, mindfulness, and humanistic management. "It's going to feel like a made-up conversation that people don't necessarily want to tune in to."

F. The most interesting people aren't those who've gone on some Eat, Pray, Love journey to find themselves. Instead, Pirson says, they're those who examine the ordinary. "Often the 'boring things' may not be boring at all. Maybe they are actually little miracles," he says. Share your observations about the world around you—interesting stories you heard or things you noticed—and you may be surprised by the universal connection they inspire.

G. This is essentially how Jessica Hagy starts her day. The author of *How to Be Interesting: An Instruction Manual*, Hagy spends a lot of time thinking about what's interesting to her. People who are interesting are persistently curious, she says.

H. Think about the everyday things around you and ask questions about them. What is that roadside monument I see on my way to work every day? Who built that interesting building in my city? What nearby attractions haven't I visited? Why do people do things that way? Use what you find to ask more questions and learn more about the world around you "Having that sort of curiosity is almost like a protective gear from getting into boredom," she says. And when you find things that are truly interesting to you share them.

I. Television veteran Audrey Morrissey, executive producer of NBC's The Voice, is always looking for what will make a person or story interesting to viewers: It's usually a matter of individuality. "Having a strong point of view signature style or being a super-enthusiast in a particular field makes someone interesting," she says. That means embracing what is truly interesting or unique about yourself. "Many people are 'not boring' in the way that they can carry a conversation or can be good at a social gathering, etc. To be interesting means that you have lived life, taken risks, traveled sought out experience to learn for yourself and share with others," she says.

J. Of course, it's possible to be a fountain of knowledge and a boring person, says public relations consultant Andrea Pass. Paying attention to the listener is an important part of having a conversation that's interesting to both parties. Talking on and on about what's interesting to you isn't going to make you an interesting person, she says.

K. "If the listener is not paying attention, it's your sign to shorten the story or change direction. Make sure to bring the audience into the conversation so that it is not one-sided," Pass says. Be a better listener yourself and give others opportunities to participate in the conversation by inviting them with questions or requests to share their own experiences or thoughts. (e.g. "Now, tell me about your favorite book," or "Have you ever been to that attraction?") Questions are a powerful tool, especially when they encourage others to disclose information about themselves. A 2012 study from the University of California, Santa Barbara found that roughly 40% of the time we are talking, we're disclosing subjective information about our experience. And when doing so, our brains are more engaged. So one strategy to leave others with the impression that you're a sparkling conversation partner is to get others to talk about themselves.

L. Being relatable is also essential, Morrissey says. "The best entertainment and storytelling comes from people who are relatable—those who don't shy away from opening up but freely share who they are and what they care about. These are the people viewers most relate to and find interesting. Being authentic, honest and vulnerable is always interesting."

M. I have now come to realize that being boring, in actuality, is not only about who you are as a person, but also how you present yourself. No matter what make sure you are having fun in life. Because when you are enjoying, people around you will begin to enjoy as well. Show some interest in them and they will definitely show some in you. If you are a very reserved person, this could be a little difficult at first. But with a little effort, you can definitely improve.

1. Pirson claims that some ordinary things may often prove to be miraculously interesting.
2. To make a conversation interesting, it is important that you listen to the other party attentively.
3. A person who is unable to stimulate others' curiosity or make their life enjoyable may appear somewhat boring.
4. Interesting people usually possess certain unique qualities, according to a TV program producer.
5. Be interested in others and they are sure to be interested in you.
6. The author considers himself usually good at conducting conversations.

7. Interesting people are always full of curiosity.
8. Falling into a routine can turn a person into an utter bore.
9. One strategy to be a good conversationalist is to motivate your partner to tell their own stories.
10. Interesting as it might appear, a made-up conversation will probably turn out to be dull.

Unit 9 Aviation Economy

Preview

The practice of overbooking, where airlines sell more tickets than available seats, is a strategy to increase profits and optimize resources. Airlines use statistical calculations based on historical data to estimate the number of passengers who will show up for a flight. By considering probabilities and associated revenues, airlines determine the optimal number of extra tickets to sell. Overbooking can significantly impact airline revenue but raises ethical concerns about selling the same resource to multiple people. Striking a balance between maximizing revenue and providing a satisfactory experience for passengers is crucial.

航空基础英语

Listening

Directions: *Fill in the blanks with the words that you hear. Find them in the table below and write down the corresponding letters.*

Overbooking is a common practice among airlines, which involves selling more tickets than the actual number of seats available. Despite its controversial nature, overbooking serves a purpose for airlines. It allows them to __1__ their resources and __2__ their profits.

Firstly, airlines understand that not all passengers will show up for their flights. Through years of data collection, they have developed __3__ models to predict the __4__ of each individual customer arriving on time. Armed with this information, airlines carefully calculate the number of tickets to sell.

Selling too few tickets would result in wasted seats and potential __5__ loss. On the other hand, selling too many tickets can lead to the inconvenience of passengers being bumped off their flights. Therefore, airlines must strike a __6__ balance to ensure optimal revenue and __7__ satisfaction.

To achieve this, airlines analyze __8__ data and utilize __9__ statistical models. These tools help them assess the likelihood of different outcomes, enabling them to make informed decisions about the number of tickets to sell.

While overbooking can cause passenger inconvenience for passengers who are bumped off their flights, airlines view it as a __10__ strategy to maximize their profits. It requires careful planning, consideration of customer behavior, and the potential costs involved.

| A. optimize | B. probability | C. statistical | D. customer | E. revenue |
| F. delicate | G. historical | H. binomial | I. maximize | J. necessary |

Reading

Why Do Airlines Sell Too Many Tickets?

Overbooking is a common practice among businesses and institutions to **maximize** profits and **optimize** resources. While it can be frustrating for customers, overbooking occurs because companies **anticipate no-shows** and sell more tickets than their actual **capacity**.

Airlines serve as a classic example of overbooking, with approximately 50,000 passengers being **bumped** off flights each year. However, this **figure** doesn't come as a surprise to airlines themselves, as they employ **statistical** data to determine the **optimal** number of tickets to sell. **Striking the right balance** is crucial, as selling too few seats would result in wasted opportunities, while selling too many would lead to **penalties**, **compensation**, and dissatisfied customers.

To calculate the appropriate number of tickets, airlines rely on years' worth of data regarding customer no-show rates for specific flights. For instance, they might determine that on a particular route, the probability of an individual customer showing up on time is 90 percent. **Assuming** each customer

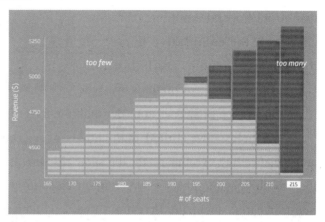

travels individually rather than in groups, if there are 180 seats available and 180 tickets are sold, the most likely outcome is that 162 passengers will board. However, **variations** are possible, and the **probability** for each **value** is determined by a **binomial distribution**, which **peaks** at the most likely outcome.

Revenue plays a crucial role in this decision-making process. Airlines make money from ticket sales, but they also **incur** costs for bumping passengers. Let's consider an example where a ticket costs $250 and is non-**refundable** or **transferrable**, while the cost of bumping a passenger is $800 (note that actual amounts vary). If no extra tickets are sold, the airline would earn $45,000. However, if they sell 15 extra tickets and assume at least 15 no-shows, the revenue would increase to $48,750, representing the **best-case scenario**. On the other hand, if all passengers show up and 15 individuals get bumped, the revenue would decrease to $36,750, even lower than if only 180 tickets were sold **initially**.

Calculating the **likelihood** of each scenario is important, not just its **financial implication**s. The binomial distribution helps determine the probability of specific passenger counts. For instance, the probability of exactly 195 passengers boarding would be nearly zero percent, while the probability of exactly 184 passengers boarding is 1.11 percent, and so on. By **multiplying** these probabilities with the **corresponding** revenue and **subtracting** the sum from the earnings of selling 195 tickets, the expected revenue for selling 195 tickets can be **obtained**.

Airlines can repeat this calculation for various numbers of extra tickets to find the quantity that is likely to **yield** the highest revenue. In this example, selling 198 tickets would probably generate $48,774, nearly $4,000 more than without overbooking. Considering that airlines operate millions of flights each year, the **cumulative** impact of overbooking becomes **substantial**. Naturally, the actual calculations involve more complexity, as airlines consider multiple factors to create even more **accurate models**.

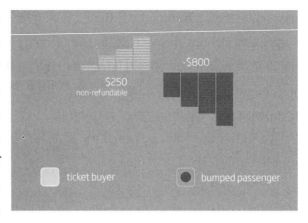

Nevertheless, the **ethical** implications of overbooking **are subject to** debate. Some argue that selling two tickets for the same resource is unethical, but it may be acceptable if the probability of a no-show is certain or close to it. However, as the **certainty** decreases, questions arise about the **threshold** that separates **practicality** from ethical concerns—whether it's 95 percent, 75 percent, or another figure.

New Words and Expressions

overbook v. /ˌəʊvəˈbʊk/	to sell more tickets on a plane or reserve more rooms in a hotel than the places available 超售（飞机座位或旅馆客房）
maximize v. /ˈmæksɪmaɪz/	to make the best use of sth. 充分利用，最大限度地利用
optimize v. /ˈɒptɪmaɪz/	to make sth. as good as it can be; to use sth. in the best possible way 使最优化，充分利用
anticipate v. /ænˈtɪsɪpeɪt/	to expect sth. 预料，预期
no-show n.	a person who is expected to be somewhere and does not come; a situation where this happens 没有如期出现的人，失约，放弃预订
capacity n. /kəˈpæsəti/	the number of things or people that a container or space can hold

	容量，容积，容纳能力
bump *v.* /bʌmp/	to move sb. from one group or position to another; to remove sb. from a group
	掉换，调出，开除
figure *n.* /ˈfɪɡjər/	a number representing a particular amount, especially one given in the official information
	数字
statistical *adj.* /stəˈtɪstɪk(ə)l/	relating to numbers obtained from analyzing information
	统计数字的，统计数据的
optimal *adj.* /ˈɒptɪməl/	best or most favorable, optimum
	最优的，最佳的
penalty *n.* /ˈpenəlti/	a punishment for breaking a law, rule or contract
	惩罚，处罚，刑罚
compensation *n.* /ˌkɒmpenˈseɪʃn/	financial or spiritual things given to make up for a damage
	赔偿，补偿金
assume *v.* /əˈsjuːm/	to think or accept that sth. is true but without having proof of it
	假定，假设，认为
variation *n.* /ˌveəriˈeɪʃn/	a change, especially in the amount or level of sth.
	（数量、水平等的）变化，变更，变异
probability *n.* /ˌprɒbəˈbɪləti/	how likely sth. is to happen
	可能性，或然性
value *n.* /ˈvæljuː/	how much sth. is worth compared with its price
	（与价格相比的）值，划算程度
binomial *n.* /baɪˈnəʊmiəl/	(mathematics) an expression that has two groups of numbers or letters, joined by the sign + or –
	二项式

peak *v.* /piːk/	to reach the highest point or value 达到高峰，达到最高值
revenue *n.* /ˈrevənjuː/	the money that a government receives from taxes or that an organization, etc. receives from its business 财政收入，税收收入，收益
incur *v.* /ɪnˈkɜː(r)/	cause, bring 引致，带来（成本、花费等）
refundable *adj.* /rɪˈfʌndəb(ə)l/	that you can return 可退还的，可偿还的
transferrable *adj.* /trænzˈfɜːrrəbl/	可转移的，可传递的，可转换的
best-case *adj.*	at the best situation 在最佳情况下的
scenario *n.* /səˈnɑːriəʊ/	a description of how things might happen in the future 设想，方案，预测
initially *adv.* /ɪˈnɪʃəli/	at the beginning 开始，最初，起初
likelihood *n.* /ˈlaɪklihʊd/	the chance of sth. happening; how likely sth. is to happen 可能，可能性
multiply *v.* /ˈmʌltɪplaɪ/	to add a number to itself a particular number of times 乘，乘以
corresponding *adj.* /ˌkɒrəˈspɒndɪŋ/	matching or connected with sth. that you have just mentioned 符合的，相应的，相关的
subtract *v.* /səbˈtrækt/	to take a number or an amount away from another number or amount 减，减去
obtain *v.* /əbˈteɪn/	to get sth., especially by making an effort （尤指经努力）获得，赢得
yield *v.* /jiːld/	to produce or provide sth., for example a profit, result or crop

Unit 9 Aviation Economy

	出产（作物），产生（收益、效益等），提供
cumulative *adj.* /ˈkjuːmjələtɪv/	having a result that increases in strength or importance each time more of sth. is added （在力量或重要性方面）聚积的，积累，渐增的
substantial *adj.* /səbˈstænʃ(ə)l/	large in amount, value or importance 大量的，价值巨大的，重大的
accurate *adj.* /ˈækjərət/	correct and true in every detail 正确无误的
model *n.* /ˈmɒdl/	a simple description of a system, used for explaining how sth. works or calculating what might happen, etc. （用于示范运作方法等的）模型
ethical *adj.* /ˈeθɪkl/	morally correct or acceptable 合乎道德的
certainty *n.* /ˈsɜːtnti/	a thing that is certain 确实的事，必然的事
threshold *n.* /ˈθreʃhəʊld/	the floor or ground at the bottom of a doorway; entrance to a room or building; the level at which sth. starts to happen or have an effect 门槛，门口，阈，界，起始点
practicality *n.* /ˌpræktɪˈkæləti/	the quality of being suitable, or likely to be successful; the real facts and circumstances rather than ideas or theories 可行性，适用性，实际事物，实际情况
strike a balance	结帐，公平处理
binomial distribution	二项分布
financial implication	财务影响
be subject to	受制于，遭受，易受影响

Genre Analysis

Genre:
The genre of this text is an informative article or a business analysis.

Communicative Moves:

Move 1 Introduction:
The article introduces the practice of overbooking and explains its purpose in maximizing profits and resource optimization.

Move 2 Explanation of Overbooking in the Airline Industry:
The article discusses how airlines overbook flights based on statistical data and customer no-show rates to find the optimal number of tickets to sell.

Move 3 Financial Considerations:
The article explores the financial aspects of overbooking, including revenue from ticket sales and costs associated with bumping passengers, using a specific example to illustrate the impact on earnings.

Move 4 Probability Calculations:
The article explains the use of binomial distribution to calculate the likelihood of different scenarios, such as the number of passengers boarding a flight, and how it helps in determining expected revenue.

Move 5 Decision-making Process:
The article describes how airlines can repeat the calculation for various numbers of extra tickets to find the quantity that is likely to generate the highest revenue.

Move 6 Cumulative Impact and Complexity:
The article highlights the cumulative impact of overbooking for airlines operating millions of flights and acknowledges the complexity involved in creating accurate models.

Move 7 Ethical Considerations:
The article briefly touches upon the ethical implications of overbooking, including the debate around its acceptability based on the probability of no-shows and the threshold that separates practicality from ethical concerns.

Communicative Purpose:
The purpose of this article is to inform readers about the practice of overbooking in the airline industry, explaining the reasons behind it and the decision-making process involved. It aims to provide insights into the financial considerations and probability calculations airlines use to determine the optimal number of tickets to sell. Additionally, the article

briefly raises the ethical concerns associated with overbooking, acknowledging that the acceptability of the practice is subject to debate.

Culture Tips

1. Respect different perspectives on overbooking: The practice of overbooking and its acceptability can vary across cultures. Recognize that opinions on this topic may differ, and it's important to respect and understand different cultural viewpoints on the matter.

2. Adapt to local customer service expectations: When interacting with airline staff or sales agents from different cultures, be aware that customer service expectations can vary. Some cultures may prioritize formal and polite interactions, while others may have a more relaxed and informal approach. Adjust your communication style accordingly to ensure effective and respectful interactions.

3. Be patient with cultural nuances: Cultural nuances can influence the decision-making process and business practices related to overbooking. If you encounter differences in how airlines handle overbooking in different countries or regions, be patient and seek to understand the underlying cultural factors that may be contributing to those practices.

4. Understand the impact of cultural values: Cultural values such as individualism, collectivism, and risk tolerance can influence how airlines approach overbooking. Take the time to learn about the cultural values that shape business decisions in different regions, as it will provide valuable insights into the practices and strategies employed by airlines.

5. Embrace cultural diversity in travel experiences: As a traveler, you will likely encounter a diverse range of cultures in your interactions with airlines. Embrace the opportunity to learn from these cultural encounters, appreciate the diversity, and adapt your behavior to show respect and understanding towards different cultural perspectives.

6. Remember, cultural tips are meant to provide guidance and promote understanding, but it's important to approach each cultural interaction with an open mind and a willingness to learn and adapt. Being respectful, patient, and adaptable will help you navigate cultural differences effectively and foster positive experiences in your travels.

Reading Exercises

I. Vocabulary Matching

Directions: *Fill in the blank with the corresponding letter of the Chinese equivalent of the following English words.*

1. sell ( )   7. overbooking ( )   13. customer ( )   19. boarding ( )
2. ticket ( )   8. passengers ( )   14. exactly ( )   20. businesses ( )
3. dollars ( )   9. revenue ( )   15. money ( )   21. calculation ( )
4. percent ( )   10. bump ( )   16. people ( )   22. extra ( )
5. airline ( )   11. probability ( )   17. actual ( )   23. find ( )
6. flight ( )   12. case ( )   18. scenario ( )   24. full ( )

A. 场景   G. 超额预定   M. 登机   S. 实际的
B. 百分比   H. 乘客   N. 确切地   T. 酒店
C. 收入   I. 票   O. 钱   U. 计算
D. 出售   J. 撞   P. 客户   V. 企业
E. 航空公司   K. 情况   Q. 航班   W. 找到
F. 美元   L. 人   R. 概率   X. 额外的
Y. 数字   Z. 完整的

II. Answer the Following Questions

1. Why do airlines engage in overbooking?
2. How do airlines determine the number of tickets to sell?
3. What role does revenue play in the decision-making process for overbooking?
4. How do airlines calculate the likelihood of different passenger counts?
5. What are the ethical implications of overbooking?

III. Multiple Choice

1. Why do airlines sell too many tickets? ( )

   A. Overbooking is a common practice among businesses and institutions to maximize profits and optimize resources.

   B. Airlines serve as a classic example of overbooking, with approximately 50,000 passengers being bumped off flights each year.

   C. To calculate the appropriate number of tickets, airlines rely on years' worth of data regarding customer no-show rates for specific flights.

Unit 9 Aviation Economy

   D. Revenue plays a crucial role in this decision-making process.
2. How do airlines determine the optimal number of tickets to sell? (　)
   A. By relying on statistical data.
   B. By inspecting the passengers in the airport.
   C. By asking the passengers.
   D. By consulting the government.
3. What is the purpose of using the binomial distribution in overbooking calculations? (　)
   A. To determine the appropriate number of tickets to sell.
   B. To calculate the financial implications of overbooking.
   C. To estimate the probability of specific passenger counts.
   D. To optimize resources and maximize profits.
4. Why do airlines engage in overbooking despite the risks involved? (　)
   A. To maximize revenue and avoid wasted opportunities.
   B. To minimize compensation and penalties.
   C. To ensure a higher likelihood of passengers showing up.
   D. To meet the demand for tickets on popular routes.
5. What is one ethical concern associated with overbooking? (　)
   A. Selling two tickets for the same resource is considered unethical.
   B. The threshold that separates practicality from ethical concerns is uncertain.
   C. Overbooking leads to dissatisfied customers and compensation claims.
   D. The probability of a no-show is subject to debate.

Translation

Directions: *Translate the following sentences.*

1. In the US, about 50,000 people get bumped off their flights due to the oversell of airlines each year.

2. The airline makes money from each ticket buyer and loses money for each person who gets bumped.

3. By repeating this calculation for various numbers of extra tickets, the airline can find the one likely to yield the highest revenue in this example.

4. If you don't sell any extra tickets, you make 45,000 dollars. If you sell 15 extras and at least 15 people are no-shows, you make 48750 dollars.

5. We can find out by using the binomial distribution in this example, the probability of exactly 195 passengers boarding is almost zero percent.

6. 尽管超售常常让客户感到愤怒，但其存在的原因是能让航空公司增加利润、优化资源。

Speaking

Student: Hi, I have some questions about why airlines sell more tickets than they have seats. Can you help me understand that?

Sales agent: Sure! Airlines sell extra tickets because they want to make the most money and use their resources efficiently. They expect some passengers to not show up for their flights, so they sell more tickets than there are seats.

Student: But what happens if everyone shows up?

Sales agent: Good question. Airlines look at past data to figure out how many tickets to sell. They use information about how often customers don't show up to calculate the right number. They want to make money but also avoid selling too many tickets and causing problems.

Student: How do they know the right number of tickets to sell?

Sales agent: Airlines study data from previous flights to make their decision. Let's say there are 180 seats available, and they know from past flights that about 90% of customers actually show up. Based on this, they can sell more tickets, expecting some people to not show up. This way, they can make money without selling more tickets than the plane can hold.

Student: So, it's about finding a balance between making money and not causing issues for passengers?

Sales agent: Exactly! Selling too few seats means missing out on money, but selling too many would lead to problems like fees, compensation, and unhappy customers. Airlines think about these things when they decide how many tickets to sell.

Student: I get it now. I guess the money part is really important?

Sales agent: Yes, it's a big factor. Airlines make money from ticket sales, but they also have to pay if they have to bump passengers because of overbooking. They look at things like ticket prices, costs of bumping passengers, and potential revenue from selling extra tickets to figure out what will make them the most money.

Student: It sounds like a lot of calculations. How do they figure out the probabilities and how much money they'll make?

Sales agent: Airlines use special calculations to figure out the chances of different things happening. They look at past data and use something called the binomial distribution. It helps them calculate the chances of different numbers of passengers showing up. By multiplying those chances with the money they'll make, they can figure out the best option.

Student: Wow, that's interesting. Does this overbooking thing really affect airlines?

Sales agent: Yes, it does. Airlines have thousands of flights every year, so overbooking adds up. But remember, the actual calculations are more complicated than what we talked about. Airlines consider many things to make their models even more accurate.

Student: Lastly, what about the ethics of all this?

Sales agent: That's a good question. People have different opinions about the ethics of overbooking. Some think it's not right to sell two tickets for the same seat. Others say it's okay if they're pretty sure some people won't show up. There's still a debate about what's acceptable and what's not.

Student: Thanks for explaining all that. It's cool to learn about how airlines handle ticket sales. I appreciate your help!

Sales agent: You're welcome! I'm glad I could help you understand. If you have any more questions, just let me know.

Writing

Directions: *Some people think that economic development should take priority over environmental protection as this strengthens the economy, provides jobs and increases people's standards of living while others think this is only short-term thinking and that mankind has a long-term responsibility to protect the environment. What is your opinion about this?*

_____
_____
_____
_____
_____
_____
_____
_____

Expansion Exercise for CET-BAND 4—Reading

Directions: *The passage is followed by some questions or unfinished statements. For each of them there are four choices marked A, B, C and D. You should decide on the best choice.*

Success was once defined as being able to stay at a company for a long time and move up the corporate ladder. The goal was to reach the top, accumulate wealth and retire to a life of ease. My father is a successful senior executive. In 35 years, he worked for only three companies.

When I started my career, things were already different. If you weren't changing companies every three or four years, you simply weren't getting ahead in your career. But back then, if you were a consultant or freelancer（自由职业者）, people would wonder what was wrong with you. They would assume you had problems getting a job.

Today, consulting or freelancing for five businesses at the same time is a badge of honor. It shows how valuable an individual is. Many companies now look to these "ultimate professionals" to solve problems their full-time teams can't. Or they save money by hiring "top-tier（顶尖的）experts" only for particular projects.

Working at home or in cafes, starting businesses of their own, and even launching business ventures that eventually may fail, all indicate "initiative" "creativity" and "adaptability", which are desirable qualities in today's workplace. Most importantly, there is a growing recognition that people who balance work and play, and who work at what they are passionate about, are more focused and productive delivering greater value to their clients.

Who are these people? They are artists, writers, programmers, providers of office services and career advice. What's needed now is a marketplace platform specifically designed to bring freelancers and clients together. Such platforms then become a place to feature the most experienced professional, and creative talent. This is where they conduct business, where a sense of community reinforces the culture and values of the gig economy （零工经济）, and where success is rewarded with good reviews that encourage more business.

Slowly but surely, these platforms create a bridge between traditional enterprises and this emerging economy. Perhaps more importantly, as the global economy continues to be disrupted by technology and other massive changes, the gig economy will itself become an engine of economic and social transformation.

1. What does the author use the example of his father to illustrate? (　　)
    A. How long people took to reach the top of their career.
    B. How people accumulated wealth in his father's time.
    C. How people viewed success in his father's time.
    D. How long people usually stayed in a company.
2. Why did people often change jobs when the author started his career? (　　)
    A. It was considered a fashion at that time.
    B. It was a way to advance in their career.
    C. It was a response to the changing job market.
    D. It was difficult to keep a job for long.
3. What does the author say about people now working for several businesses at the same time? (　　)
    A. They are often regarded as the most treasured talents.
    B. They are able to bring their potential into fuller play.
    C. They have control over their life and work schedules.
    D. They feel proud of being an outstanding problem solver.

4. What have businesses come to recognize now?　(　)

   A. Who is capable of solving problems with ease.

   B. How people can be more focused and productive.

   C. What kind of people can contribute more to them.

   D. Why some people are more passionate about work.

5. What does the author say about the gig economy?　(　)

   A. It may force companies to reform their business practice.

   B. It may soon replace the traditional economic model.

   C. It will drive technological progress on a global scale.

   D. It will bring about radical economic and social changes.

Unit 10 Aviation Trend

Preview

The C919, a locally built narrow-body commercial jet, represents China's pursuit of self-sufficiency and independence in its development. Initiated in 2008 by COMAC, the project aimed to reduce China's reliance on foreign aircraft manufacturers like Boeing and Airbus. Despite facing challenges and delays, COMAC remained committed to delivering an exceptional aircraft, leveraging technology and forming strategic partnerships to optimize performance. The C919's development exceeded the estimated budget but showcased advanced technology and craftsmanship upon its unveiling in 2015. Overcoming obstacles such as supply restrictions, the C919 took its first flight in 2017 and entered commercial service in 2023. With its advanced features and competitive capabilities, the C919 challenges the dominance of Boeing and Airbus, offering Chinese and international airlines a viable alternative and fostering healthy competition in the aviation market. Its success signifies China's determination to become a major player in aircraft manufacturing, paving the way for future advancements and bolstering its quest for self-sufficiency and independence in this critical field.

Listening

Directions: *Fill in the blanks with the words that you hear. Find them in the table below and write down the corresponding letters.*

The COMAC C919 project began in 2008 with the goal of creating an __1__ narrow-body aircraft. COMAC, a Chinese aircraft manufacturer, had __2__ plans to produce up to 2,300 C919 planes. This was to have an entire fleet of China-made passenger jets in all the __3__ operating in China. In 2011, COMAC formed a __4__ with Ryanair, a well-known Irish low-cost airline, to combine their __5__ and accelerate the development of the C919. This __6__ caught the attention of Airbus, a major player in the aviation industry, which recognized the C919's __7__ to be a strong competitor. By 2020, __8__ jet makers knew that COMAC was the company having the __9__ to crack their hegemony, hence plans about creating __10__ started.

A. potential	B. international	C. advanced	D. ambitious	E. airlines
F. muscles	G. hurdles	H. partnership	I. expertise	J. collaboration

Reading

China's C919 Takes First Flight, Posing a Challenge to Boeing and Airbus

China's resolute efforts to achieve self-sufficiency and independence in all aspects of its development has led to the creation of the C919, a locally built narrow-body commercial jet. The project, initiated in 2008 by COMAC, a Chinese aircraft  manufacturer, aimed to reduce China's dependency on foreign aircraft manufacturers like Boeing and Airbus. This move caught the attention of international players in the aviation industry, who recognized the C919's potential as a **formidable** competitor.

The development process of the C919 faced numerous challenges and experienced delays, extending the anticipated **timeframe** of completion to 15 years. Technical complexities and supply issues were among the obstacles encountered by COMAC. However, the company remained **steadfast** in its commitment to delivering an **exceptional** aircraft, continuously refining the design and manufacturing processes to optimize performance and safety. To support these efforts, COMAC **leveraged** the power of the Tianhe-2 supercomputer, known for its impressive processing capabilities, enabling precise calculations and simulations.

Recognizing the need for international **expertise** and technology, COMAC formed **strategic** partnerships with **renowned** companies such as Ryanair and Bombardier Aerospace. These collaborations covered various areas, including supply chain services, electrical systems, cockpit development, flight training, and sales and marketing support. By combining the strengths of both domestic and international **entities**, COMAC aimed to shape the future of aviation and establish itself as a major player in the global market.

Financial challenges also presented themselves during the development of the C919, with the actual cost exceeding the initial estimated budget of 58 billion yuan, reaching over 20 billion dollars. This substantial investment reflected the extensive research, development, and testing required to bring the C919 to **fruition**. However, despite these financial **constraints**, COMAC remained **resolute** in its pursuit of creating a world-class aircraft.

After years of **perseverance**, the first C919 **prototype** was **unveiled** in November 2015, showcasing the advanced technology and **meticulous craftsmanship** involved in its creation. However, further **hurdles** awaited COMAC on its path to **commercialization**. In 2020, the Trump administration imposed restrictions on the **shipment** of crucial components such as flight controls and jet engines, under the pretext of concerns of spying and **espionage**. This move significantly **hindered** COMAC's progress, as approximately 60 percent of the required parts came from American suppliers. To overcome this obstacle, COMAC had to obtain special licenses from companies like General Electric and

Honeywell, enabling access to the necessary components.

Furthermore, COMAC faced challenges in obtaining local certification from China's aviation regulator, the Civil Aviation Administration of China (CAAC), due to incomplete operations resulting from the restricted access to certain parts. Despite these **intentional** obstacles, COMAC's determination and **relentless** efforts prevailed. In 2017, the C919 took its first flight, and in 2023, it finally entered commercial service.

The COMAC C919 represents a significant challenge to the long-established **dominance** of Boeing and Airbus in the aviation industry. With its advanced technology, **fly-by-wire** flight control system, modern avionics, and improved fuel efficiency, the C919 competes directly with popular models like the Boeing 737 and Airbus A320. It has a passenger capacity ranging from approximately 168 to 190, depending on the **configuration**, and a range of approximately 4,075 to 5,555 kilometers (2,530 to 3,450 miles).

China's domestic aviation market is one of the largest and fastest-growing in the world. By introducing the C919, Chinese airlines have the opportunity to rely on a locally produced option that meets their specific needs, potentially reducing their dependence on foreign aircraft. Additionally, the C919 presents a **viable alternative** for airlines worldwide, offering them more choices and **fostering** healthy competition in the aviation market.

While breaking into an industry dominated by established players like Boeing and Airbus is undoubtedly challenging, the successful entry of the C919 into commercial service marks a significant milestone for China's aviation industry. It showcases China's determination to become a major player in aircraft manufacturing and poses a potential challenge to the **duopoly**. The success of the C919 lays the foundation for future advancements and developments in China's aviation sector, as it continues to strive for self-sufficiency and independence in this critical field.

New Words and Expressions

formidable *adj.* /ˈfɔːmɪdəbl/	impressive, powerful or very difficult, can make others feel fear and/or respect for them 可怕的，令人敬畏的，难对付的
timeframe *n.* /ˈtaɪmfreɪm/	schedule, timetable 时间表
steadfast *adj.* /ˈstedfɑːst/	not changing in your attitudes or aims 坚定的，不动摇的
▲exceptional *adj.* /ɪkˈsepʃənl/	unusually good 杰出的，优秀的，卓越的
leverage *v.* /ˈliːvərɪdʒ/	use ... to maximum advantage 最大限度的利用，最优化适用
n.	the ability to influence what people do; the act of using a lever to open or lift sth.; the force used to do this 影响力，杠杆作用，杠杆效力
expertise *n.* /ˌekspɜːˈtiːz/	expert knowledge or skill in a particular subject, activity or job 专门知识，专门技能，专长
strategic *adj.* /strəˈtiːdʒɪk/	done as part of a plan that is meant to achieve a particular purpose or to gain an advantage 根据全局而安排的，战略性的
renowned *adj.* /rɪˈnaʊnd/	famous and respected 有名的，闻名的，受尊敬的
entity *n.* /ˈentəti/	something that exists separately from other things and has its own identity 独立存在物，实体
fruition *n.* /fruˈɪʃn/	the successful result of a plan, a process or an activity （计划、过程或活动的）完成，实现，取得成果
constraint *n* /kənˈstreɪnt/	a thing that limits or restricts sth., or your freedom to do sth.

		限制，限定，约束
▲resolute *adj.* /ˈrezəluːt/		having or showing great determination
		坚决的，有决心的
perseverance *n.* /ˌpɜːsɪˈvɪərəns/		the quality of continuing to try to achieve a particular aim despite difficulties
		毅力，韧性，不屈不挠的精神
prototype *n.* /ˈprəʊtətaɪp/		the first design of sth. from which other forms are copied or developed
		原型，雏形，最初形态
unveil *v.* /ˌʌnˈveɪl/		to show or introduce a new plan, product, etc. to the public for the first time
		（首次）展示，介绍，推出，将……公之于众
meticulous *adj.* /məˈtɪkjələs/		paying careful attention to every detail
		细心的，小心翼翼的
craftsmanship *n.* /ˈkrɑːftsmənʃɪp/		the level of skill shown by sb. in making sth. beautiful with their hands; the quality of such works
		手艺，技艺，精工细作
hurdle *n.* /ˈhɜːdl/		a problem or difficulty that must be solved or dealt with before you can achieve sth.
		难关，障碍
commercialization *n.* /kəˌmɜːrʃələˈzeɪʃn/		商品化，商业化
shipment *n.* /ˈʃɪpmənt/		the process of sending goods from one place to another
		运输，运送，装运
espionage *n.* /ˈespiənɑːʒ/		the activity of secretly getting important political or military information about another country or of finding out another company's secrets by using spies
		间谍活动，谍报活动，刺探活动
hinder *v.* /ˈhɪndə(r)/		to make it difficult for sb. to do sth. or sth. to happen
		阻碍，妨碍，阻挡
intentional *adj.* /ɪnˈtenʃənl/		done deliberately

	故意的，有意的，存心的
relentless *adj.* /rɪˈlentləs/	not stopping or getting less strong
	不停的，持续强烈的，不减弱的
▲dominance *n.* /ˈdɒmɪnəns/	the fact that some people or things are more powerful, successful, or important than others
	优势，统治地位
fly-by-wire *n.*	aircraft control through systems operated by electronic circuits rather than mechanical rods
	电传操控
configuration *n.* /kənˌfɪɡəˈreɪʃn/	an arrangement of the parts of sth. or a group of things; the form or shape that this arrangement produces
	布局，结构，构造，格局，形状
viable *adj.* /ˈvaɪəbl/	that can be done; that will be successful
	可实施的，切实可行的
alternative *adj.* /ɔːlˈtɜːnətɪv/	a thing that you can choose to do or have out of two or more possibilities
	可供选择的事物
foster *v.* /ˈfɒstə(r)/	to encourage sth. to develop
	促进，助长，培养，鼓励
duopoly *n.* /djuːˈɒpəli/	a right to trade in a particular product or service, held by only two companies or organizations
	（商品或服务的）两强垄断

Genre Analysis

Genre:

The passage can be categorized as an informative article or news report, providing detailed information about China's C919 aircraft, its development process, challenges, partnerships, and its potential impact on the aviation industry.

Communicative Moves:

Move 1 Introducing the C919 and its Purpose:

The passage begins by introducing the C919, highlighting its significance as a locally built commercial jet and China's aim to reduce dependency on foreign aircraft manufacturers.

Move 2 Describing the Development Challenges:

The passage discusses the challenges faced by COMAC during the development of the C919, including technical complexities, supply issues, and financial constraints. It emphasizes the company's determination to overcome these obstacles.

Move 3 Highlighting Strategic Partnerships:

The passage mentions the strategic partnerships formed by COMAC with renowned companies in various areas, demonstrating the importance of international expertise and collaboration in the C919's development.

Move 4 Addressing Regulatory and Certification Challenges:

The passage discusses the restrictions imposed by the US government on the shipment of crucial components and the challenges faced in obtaining local certification from China's aviation regulator. It emphasizes COMAC's perseverance in overcoming these obstacles.

Move 5 Presenting the Features and Competitiveness of the C919:

The passage describes the advanced technology, fly-by-wire flight control system, modern avionics, improved fuel efficiency, passenger capacity, and range of the C919, highlighting its direct competition with Boeing and Airbus models.

Move 6 Discussing the Market Impact:

The passage highlights the potential impact of the C919 on China's domestic aviation market, reducing dependence on foreign aircraft for Chinese airlines and offering more choices and competition for airlines worldwide.

Move 7 Reflecting on the Milestone and Future Outlook:

The passage concludes by acknowledging the milestone achieved with the C919's entry into commercial service, emphasizing China's determination to become a major player in aircraft manufacturing and its aspirations for self-sufficiency and independence in the aviation sector.

Communicative Purpose:

The purpose of the passage is to inform readers about the development, challenges, partnerships, and potential market impact of China's C919 aircraft. It aims to highlight China's efforts to establish itself as a major player in the aviation industry and challenge the dominance of Boeing and Airbus. The passage also emphasizes China's ambition for self-sufficiency and independence in aircraft manufacturing.

Culture Tips

1. Determination and perseverance: The development of the C919 exemplifies China's commitment to achieving its goals. Chinese culture places great emphasis on persistence and resilience in the face of challenges. When working with individuals from Chinese backgrounds, it is important to recognize and respect their determination and dedication to their objectives.
2. Collaboration and partnerships: COMAC formed strategic partnerships with renowned international companies during the development of the C919. Collaborative efforts and building relationships play a significant role in Chinese business culture. When interacting with Chinese counterparts, consider fostering partnerships and finding common ground for mutual benefit.
3. Respect for technology and innovation: The use of advanced technology, such as the Tianhe-2 supercomputer, demonstrates China's emphasis on technological advancements. Chinese culture highly values innovation, scientific progress, and the application of technology in various fields. Acknowledging and appreciating China's technological achievements can help establish positive connections with individuals from Chinese backgrounds.
4. Face and saving face: Chinese culture places importance on "face", which refers to preserving one's dignity, reputation, and social standing. It is crucial to be mindful of this concept when providing feedback or discussing challenges. Maintaining harmony and avoiding direct confrontation can help foster positive relationships and save face for everyone involved.
5. Long-term perspective: The development of the C919 spanned over 15 years, showcasing China's long-term planning and vision. Chinese culture often takes a broader perspective and values long-term goals over immediate results. When engaging with individuals from Chinese backgrounds, it can be beneficial to consider long-term partnerships and objectives.
6. Respect for local regulations: COMAC faced challenges in obtaining local certification from China's aviation regulator. Respecting and adhering to local regulations is essential when doing business in China. Familiarize yourself with the specific regulatory frameworks and cultural norms to ensure compliance and build trust with Chinese partners.

Reading Exercises

I. Vocabulary Matching

Directions: *Fill in the blank with the corresponding letter of the Chinese equivalent of the following English words.*

1. adventure ( )   8. emerged ( )   15. cancel ( )   22. despite ( )
2. crisis ( )   9. deregulate ( )   16. schedule ( )   23. proposal ( )
3. innovative ( )   10. privilege ( )   17. progress ( )   24. rise ( )
4. pandemic ( )   11. propose ( )   18. demand ( )   25. skeptical ( )
5. frequent ( )   12. academic ( )   19. infrastructure ( )   26. original ( )
6. domestic ( )   13. convince ( )   20. available ( )
7. exclusive ( )   14. revolution ( )   21. produce ( )

A. 冒险，冒险经历　　　　J. 出现，浮现　　　　S. 要求，需求
B. 取消，撤销　　　　　　K. 频繁的，经常发生的　T. 撤销管制
C. 流行病，瘟疫　　　　　L. 基础设施　　　　　　U. 提议，建议
D. 危机，危急关头　　　　M. 专用的，专有的　　　V. 升起，增加
E. 特殊利益，优惠待遇　　N. 使确信，使相信　　　W. 即使，尽管
F. 革新的，创新的　　　　O. 革命　　　　　　　　X. 增加，提高
G. 本国的，国内的　　　　P. 学业的，教学的　　　Y. 怀疑的
H. 生产，制造　　　　　　Q. 原来的，起初的　　　Z. 可获得的
I. 工作计划　　　　　　　R. 进步，进展

II. Answer the Following Questions

1. When did China's development of the C919 passenger jet begin?
2. What challenges did COMAC face during the C919 project?
3. When did the first C919 prototype get published?
4. How does the C919 aim to compete with Boeing and Airbus?
5. What is China's goal regarding the production of the C919's components?

III. Multiple Choice

1. When did the C919 embark on its inaugural commercial passenger service? ( )

    A. 2008.
    B. 2015.
    C. 2023.

D. 2020.

2. What were some of the challenges faced by COMAC during the C919 project? (　)
   A. Financial difficulties and delays.
   B. Political interference and restrictions on component shipments.
   C. Technical complexities and supply issues.
   D. All of the above.

3. Which international airline partnered with COMAC to accelerate the development of the C919? (　)
   A. Boeing.
   B. Airbus.
   C. Ryanair.
   D. Bombardier Aerospace.

4. What impact could the success of the C919 have on Boeing and Airbus? (　)
   A. It could reduce China's reliance on foreign technology.
   B. It could compete with planes like the Boeing 737 Max and Airbus A320.
   C. It could potentially affect their market share.
   D. All of the above

5. What is China's ultimate goal regarding the production of the C919's components? (　)
   A. To rely entirely on domestic production for all components.
   B. To import components from other countries.
   C. To reduce the cost of producing the C919.
   D. To develop advanced technology for the C919.

Translation

Directions: *Translate the following sentences into Chinese.*

1. Despite facing technical complexities and supply issues, COMAC remained committed to delivering an exceptional aircraft.

   _____

   _____

2. The C919 project, which began in 2008, has encountered numerous challenges along the way, but it has persevered and made significant progress.

   _____

   _____

3. China's ambition extends beyond producing its own passenger jet; it aims to achieve self-sufficiency in manufacturing all components of the C919.
   _____
   _____

4. The C919's maiden flight marked a significant milestone in China's aviation industry, showcasing its technological advancements and determination to compete on the global stage.
   _____
   _____

5. China's pursuit of producing and exporting its own passenger jets poses a potential challenge to the dominance of established manufacturers such as Boeing and Airbus.
   _____
   _____

6. COMAC's collaboration with international companies like Bombardier Aerospace has contributed to the enhancement of the C919's capabilities and overall quality.
   _____
   _____

7. The successful development and commercialization of the C919 could potentially break the duopoly of Boeing and Airbus in the global aviation market.
   _____
   _____

8. The C919's competitive pricing, combined with advanced technology and solid performance, positions it as a formidable contender in the global aviation market.
   _____
   _____

Speaking

Student: Hi, I've been reading about China's C919, and it seems like a big challenge for Boeing and Airbus. Can you tell me more about its development and the challenges faced by COMAC?

Aviation expert: Sure! COMAC started developing the C919 in 2008 to reduce China's reliance on foreign aircraft manufacturers. The project faced delays and technical issues, but COMAC remained committed. They partnered with international companies and used advanced technology to improve the aircraft. The cost exceeded the initial budget, but they persisted.

After many years, the first C919 prototype was unveiled in 2015. However, there were more challenges. The US restricted the shipment of crucial parts, and COMAC had to get special licenses. They also faced certification delays in China. Despite these obstacles, the C919 made its first flight in 2017 and entered commercial service in 2023.

Student: That's impressive! How does the C919 compete with Boeing and Airbus?

Aviation expert: The C919 directly competes with Boeing's 737 and Airbus's A320. It has advanced technology, modern avionics, and better fuel efficiency. It can carry around 168 to 190 passengers and fly about 4,075 to 5,555 kilometers. Chinese airlines can rely on the C919, reducing their dependence on foreign planes. It also provides more choices for airlines worldwide and promotes competition in the aviation market.

Student: Thanks for explaining. It's exciting to see China's progress in aircraft manufacturing.

Aviation expert: You're welcome! China's advancements are indeed fascinating. The C919 is a significant milestone for them, and it will be interesting to see how it shapes the industry. Let me know if you have any more questions!

Writing

Directions: *Suppose your university is conducting a survey to collect students' opinions of online classes. You are to write a response to the survey about their advantages and disadvantages, and what improvements can be made. You will have 30 minutes for the task. You should write at least 120 words but no more than 180 words.*

Expansion Exercise for CET-BAND 4—Translation

　　大运河 (Grand Canal) 是世界上最长的人工河，北起北京，南至杭州。它是中国历史上最宏伟的工程之一。大运河始建于公元前 4 世纪，公元 13 世纪末建成。修建之初是为了运输粮食，后来也用于运输其他商品。大运河沿线区域逐渐发展为中国的工商业中心。长久以来，大运河对中国经济发展发挥了重要作用，有力地促进了南北地区之间的人员往来和文化交流。

Unit 11 Aviation Energy

Preview

The passage discusses the possibility of flying without emissions using hydrogen-powered planes and the challenges involved. The aviation industry's carbon emissions are a significant contributor to the climate crisis, and finding eco-friendly alternatives is crucial. While several options have been proposed, such as biofuels and electric aircraft, they have limitations or have not delivered significant emissions reductions. Now, the focus is on hydrogen as a potential solution due to its versatility and potential for environmentally friendly production. However, challenges remain in scaling up green hydrogen production and developing hydrogen fuel cells for aviation. The passage emphasizes the need for investments, technological advancements, and supportive infrastructure to make these alternatives practical. In the meantime, adopting sustainable practices and offsetting emissions can help reduce the environmental impact of air travel. Achieving a greener future for aviation will require a combination of approaches and ongoing solution development.

Listening

Directions: *Fill in the blanks with the words that you hear. Find them in the table below and write down the corresponding letters.*

Now, __1__ is supposed to save the planet. "Our belief in what hydrogen represents is most __2__." Why? They say hydrogen has at least three striking features. Number one: It's a versatile energy carrier and can __3__ aircraft with high energy demand where batteries would not be a viable alternative. Hydrogen can be produced using renewable energy, meaning the production can be environmentally __4__. This so-called green hydrogen is what powers the hype. Number two: It can be __5__ and transported efficiently, unlike electricity. And number three: When hydrogen is used in a fuel cell, the only byproduct is __6__ vapor. No harmful emissions. Sounds great, right? But there are challenges. The production of hydrogen requires a lot of __7__, and currently, most hydrogen is produced using fossil fuels, which defeats the purpose. However, there are ongoing efforts to scale up the production of green hydrogen using __8__ sources. The aviation industry is exploring the use of hydrogen fuel cells in aircraft, but it's still in the early stages. The technology needs to be developed further, and __9__ for hydrogen refueling would need to be established. So while hydrogen holds promise, it's not a silver bullet __10__ just yet.

| A. water | B. pivotal | C. stored | D. renewable | E. power |
| F. solution | G. infrastructure | H. hydrogen | I. friendly | J. energy |

Reading

Will We Soon Be Able to Fly Without Emissions Using Hydrogen-powered Planes?

Can we achieve **emission**-free flying in the near future? The days of flying being a joyous adventure seem distant when we consider the **immense** amount of **fossil fuels** burned to **propel** us into the sky. Aviation has become a major contributor to the **climate crisis**, with its **carbon footprint** posing a significant threat. However, **amidst** the concerns, there is **optimism** and a **pursuit** of **innovative** ideas within the industry to reduce emissions. One **promising avenue** is the utilization of hydrogen as a fuel source. Will hydrogen-powered planes soon offer an **eco-friendly** solution? Let's delve deeper into the **hype** surrounding green hydrogen and the challenges that lie ahead.

Before the pandemic, over 20,000 planes were operating worldwide, transporting a **staggering** 4.5 billion passengers in 2019. The majority of these travelers were frequent flyers on domestic flights in the US and China, accounting for a quarter of all aviation emissions. Astonishingly, a mere

Airbus Aims to Fly First Zero-Emission Plane by 2035 - thinktourism

fraction of the global population bears the **brunt** of these emissions, while the majority have never experienced air travel. In 2019, direct CO_2 emissions from aviation accounted for at least 2% of global emissions, making it the most **energy-intensive** human activity. While options like electric cars or **dietary** changes can contribute to reducing emissions in other sectors, long-distance air travel currently lacks a viable, environmentally friendly alternative.

To understand how we arrived at this **dilemma**, let's **rewind** to the early days of aviation. The **allure** of a connected world and the luxurious experience of flying made it an **exclusive** activity. However, with the **deregulation** of the market, low-cost carriers emerged, **catering** to a growing demand for air travel. Aviation transformed from a **privilege** for a few into a service for many. Today, even populations previously unable to access air travel are **thirsting** for the freedom and **connectivity** it brings. This has led to a significant increase in the number of people flying, particularly in Asia, where a hundred million individuals are expected to fly for the first time in a single year.

The COVID-19 pandemic temporarily halted air travel, but the industry is **poised** for a comeback. Experts predict a return to pre-pandemic traffic levels by the end of 2024. However, as the desire to fly remains strong, the environmental impact of fossil fuel emissions cannot be ignored. Thankfully, the industry has proposed numerous alternative technologies over the years, **fueling** hope for a greener future. These include concepts such as **zeppelins**, **blended-wing aircraft**, solar power, and **biofuels** made from **algae**, **sugarcane**, and other sustainable sources. Biofuels, in particular, have shown promise in reducing CO_2 emissions, but their limited availability remains a challenge. Despite being marketed as a leading airline in biofuels, KLM's actual usage accounted for a mere 0.18% of their total fuel consumption in 2019, leading to legal action for misleading marketing claims.

Another highly **anticipated** solution was electric aircraft, which generated considerable hype. However, the launch of a fully electric-powered aircraft, the E-Fan 1, was ultimately canceled, and significant emissions reductions have yet to **materialize**.

EU To Fly Hydrogen-Powered Planes By 2035; JetBlue 1st Airline To Use ...

Now, the focus has shifted to hydrogen as a potential **savior** for the planet. Hydrogen possesses three notable qualities: **versatility** as an energy carrier that can power high-demand aircraft, the potential for environmentally friendly production using **renewable** energy (green hydrogen), and the emission of only water vapor when used in fuel **cells**. However, challenges **persist**. Currently, the production of hydrogen relies heavily on fossil fuels, **undermining** its environmental benefits. **Scaling** up the production of green hydrogen using renewable sources is an **ongoing** effort. In aviation, the exploration of hydrogen fuel **cells** is still in its early stages, requiring further technological development and the establishment of hydrogen refueling **infrastructure**. While hydrogen holds promise, it is not a **one-size-fits-all** solution at present.

In summary, the aviation industry recognizes the **imperative** to reduce its carbon footprint and is actively exploring alternatives such as biofuels and hydrogen. However, significant investments, technological advancements, and supportive infrastructure are needed to **translate** these alternatives into practical solutions. In the meantime, adopting **sustainable** practices like improving fuel efficiency, **optimizing** flight routes, and **offsetting** emissions through carbon offset programs can help **mitigate** the environmental impact of air travel. Ultimately, a combination of approaches, including technological innovations and changes in consumer behavior, will be necessary to achieve a more eco-friendly future of aviation. Though the challenges are complex, increasing awareness and ongoing solution development offer hope for a greener path forward.

New Words and Expressions

hydrogen *n.* /ˈhaɪdrədʒən/   a chemical element, or a gas that is the lightest of all the elements
氢，氢气

▲emission *n.* /ɪˈmɪʃn/   the production or sending out of light, heat, gas, etc.

Unit 11 Aviation Energy

	（光、热、气等的）发出，射出，排放
▲immense *adj.* /ɪˈmens/	extremely large or great
	极大的，巨大的
propel *v.* /prəˈpel/	to move, drive or push sth. forward or in a particular direction
	推动，驱动，推进
amidst *perp.* /əˈmɪdst/	the same as sth.
	在……之中，在……气氛下，四周是（等于 amid）
optimism *n.* /ˈɒptɪmɪzəm/	a feeling that good things will happen and that sth. will be successful; the tendency to have this feeling
	乐观，乐观主义
▲pursuit *n.* /pəˈsjuːt/	the act of looking for or trying to find sth.
	追求，寻找
innovative *adj.* /ˈɪnəveɪtɪv/	introducing or using new ideas, ways of doing sth., etc.
	引进新思想的，采用新方法的，革新的，创新的
promising *adj.* /ˈprɒmɪsɪŋ/	showing signs of being good or successful
	有希望的，有前途的，有出息的
avenue *n.* /ˈævənjuː/	a choice or way of making progress towards sth.
	选择，途径，手段
eco-friendly *adj.*	not harmful to the environment
	对环境无害的，环保的
hype *n.* /haɪp/	advertisements and discussion on television, radio, etc. telling the public about a product and about how good or important it is
	促销广告，过度宣传
staggering *adj.* /ˈstæɡərɪŋ/	so great, shocking or surprising that it is difficult to believe
	令人难以相信的
fraction *n.* /ˈfrækʃn/	a small part or amount of sth.
	小部分，少量，一点儿

brunt *n.* /brʌnt/	to receive the main force of sth. unpleasant
	承受某事的主要压力，首当其冲
energy-intensive *adj.*	consuming much energy
	耗能的，能源密集型的
dietary *adj.* /ˈdaɪətərɪ/	sth. to describe anything that concerns a person's diet.
	饮食的
dilemma *n.* /dɪˈlemə/	a situation which makes problems, often one in which you have to make a very difficult choice between things of equal importance
	（进退两难的）窘境，困境
rewind *v.* /ˌriːˈwaɪnd/	to make a tape in a cassette player, etc. go backwards
	重绕（磁带等），倒带，倒片
allure *n.* /əˈlʊə(r)/	the quality of being attractive and exciting
	诱惑力，引诱力，吸引力
▲exclusive *adj.* /ɪkˈskluːsɪv/	only to be used by one particular person or group; only given to one particular person or group
	（个人或集体）专用的，专有的，独有的，独占的
deregulation *n.* /diːˌreɡjʊˈleɪʃən/	the removal of controls and restrictions in a particular area of business or trade.
	(在商贸领域)解除管制
cater *v.* /ˈkeɪtə(r)/	to provide the things that a particular type or person wants, especially things that you do not approve of
	满足需要，迎合
▲**privilege** *adj.* /ˈprɪvəlɪdʒ/	a special right or advantage that a particular person or group of people has
	特殊利益，优惠待遇
thirst *v.* /θɜːst/	to feel a strong desire for sth.
	渴望，渴求
connectivity *n.* /ˌkɒnekˈtɪvɪti/	the state of being connected or the degree to which two things are connected

Unit 11 Aviation Energy

连接（度），联结（度）

poise *v.* /pɔɪz/	(be poised) (of a person or organization) be ready and prepared to do sth.
	准备做某事
fuel *v.* /ˈfjuːəl/	to increase sth.; to make sth. stronger
	增加，加强，刺激
zeppelin *n.* /ˈzepəlɪn/	a German type of large airship
	齐柏林飞艇（源自德国的大型飞艇）
biofuel *n.* /ˈbaɪəʊˌfjʊəl/	a gas, liquid, or solid from natural sources such as plants that is used as a fuel.
	生物燃料
algae *n.* /ˈældʒiː/	very simple plants with no real leaves, stems or roots that grow in or near water, including seaweed
	藻，海藻
sugarcane *n.* /ˈʃʊɡərken/	a tall tropical plant with thick stems from which sugar is made
	甘蔗
▲anticipated *adj.* /ænˈtɪsɪˌpeɪtɪd/	if an event, especially a cultural event, is eagerly anticipated, people expect that it will be very good, exciting, or interesting.
	受期盼的
materialize *v.* /məˈtɪəriəlaɪz/	take place or start to exist as expected or planned
	实现，发生，成为现实
savior *n.* /ˈseɪvjə(r)/	a person who rescues sb./sth. from a dangerous or difficult situation
	救助者，拯救者，救星
versatility *n.* /ˌvɜːrsəˈtɪləti/	having a wide variety of skills
	多功能性，多才多艺，用途广泛
▲renewable *adj.* /rɪˈnjuːəbl/	(of energy and natural resources) that is replaced naturally or controlled carefully and can therefore be used without the risk of finishing it all

171

		可更新的，可再生的，可恢复的
cell	*n.* /sel/	a device for producing an electric current, for example by the action of chemicals or light
		电池
persist	*v.* /pəˈsɪst/	to continue to exist
		维持，保持，持续存在
▲undermine	*v.* /ˌʌndəˈmaɪn/	to make sth., especially sb.'s confidence or authority, gradually weaker or less effective
		逐渐削弱（信心、权威等），使逐步减少效力
scale	*v.* /skeɪl/	(~ sth. up) to increase the size or number of sth.
		增大，扩大（规模或数量）
ongoing	*adj.* /ˈɒŋɡəʊɪŋ/	continuing to exist or develop
		持续存在的，仍在进行的，不断发展的
infrastructure	*n.* /ˈɪnfrəstrʌktʃə(r)/	the basic systems and services that are necessary for a country or an organization to run smoothly, for example buildings, transport and water and power supplies
		（国家或机构的）基础设施，基础建设
one-size-fits-all	*adj.*	designed to be suitable for a wide range of situations or needs
		通用的，一体适用的
imperative	*adj.* /ɪmˈperətɪv/	very important and needs immediate attention or action
		重要紧急的，迫切的，急需处理的
translate	*v.* /trænsˈleɪt/	to change sth., or to be changed, into a different form
		（使）转变，变为
▲sustainable	*adj.* /səˈsteɪnəbl/	involving the use of natural products and energy in a way that does not harm the environment; that can continue or be continued for a long time
		（对自然资源和能源的利用）不破坏生态平衡的，

	合理利用的，可持续的
optimize *v.* /ˈɒptɪmaɪz/	to make sth. as good as it can be; to use sth. in the best possible way
	使最优化，充分利用
offset *v.* /ˈɒfset/	to use one cost, payment or situation in order to cancel or reduce the effect of another
	抵消，弥补，补偿
mitigate *v.* /ˈmɪtɪgeɪt/	to make sth. less harmful, serious, etc.
	减轻，缓和
fossil fuel	化石燃料（如煤或石油）
climate crisis	气候危机
blended-wing aircraft	翼身融合飞机
carbon footprint	碳足迹，碳排放量

Genre Analysis

Genre:

The passage can be categorized as an informative article or opinion piece discussing the potential for emission-free flying using hydrogen-powered planes.

Communicative Moves:

Move 1 Setting the Context:

*The pass*age begins by highlighting the environmental impact of aviation and the need for emission-free flying. It establishes the current dilemma and the lack of viable alternatives for reducing emissions in long-distance air travel.

Move 2 Exploring the History and Demand for Air Travel:

The passage delves into the history of aviation, from its exclusive nature to the democratization of air travel. It emphasizes the increasing demand for flying, particularly in Asia, and the expectation of a post-pandemic resurgence in air travel.

Move 3 Discussing Previous Alternative Technologies:

The passage mentions various alternative technologies proposed in the aviation industry, such as zeppelins, blended-wing aircraft, solar power, and biofuels. It acknowledges the limitations and challenges faced by these technologies, including limited availability and unfulfilled emissions reductions.

Move 4 Shifting Focus to Hydrogen as a Potential Solution:

The passage introduces hydrogen as a potential savior for emission-free flying. It highlights hydrogen's versatility as an energy carrier, the potential for environmentally friendly production using renewable energy (green hydrogen), and its emission of only water vapor when used in fuel cells.

Move 5 Addressing Challenges and Current Limitations:

The passage discusses the challenges in hydrogen production, which currently relies heavily on fossil fuels. It emphasizes the need to scale up the production of green hydrogen and the technological development required for hydrogen fuel cells. It acknowledges that hydrogen is not a fully developed solution at present.

Move 6 Recognizing the Industry's Efforts and Necessary Steps:

The passage acknowledges the aviation industry's recognition of the need to reduce its carbon footprint and the exploration of alternative technologies. It emphasizes the need for significant investments, technological advancements, and supportive infrastructure to translate these alternatives into practical solutions.

Move 7 Highlighting Current Sustainable Practices:

The passage suggests current sustainable practices in aviation, such as improving fuel efficiency, optimizing flight routes, and offsetting emissions through carbon offset programs, as interim measures to mitigate environmental impact.

Move 8 Emphasizing the Need for a Combination of Approaches:

The passage concludes by stating that a combination of approaches, including technological innovations and changes in consumer behavior, will be necessary to achieve a more eco-friendly future of aviation. It highlights increasing awareness and ongoing solution development as sources of hope for a greener path forward.

Communicative Purpose:

The purpose of the passage is to inform readers about the potential for emission-free flying using hydrogen-powered planes. It discusses the current challenges and limitations in achieving this goal while highlighting the previous alternative technologies explored in the aviation industry. The passage aims to emphasize the need for investments, technological advancements, and supportive infrastructure to make hydrogen-powered planes and other alternatives practical solutions. It also emphasizes the importance of sustainable practices and a combination of approaches to mitigate the environmental impact of air travel.

Culture Tips

1. Stay informed: Keep yourself updated on the latest developments in sustainable aviation

technologies and practices. Stay informed about the challenges, advancements, and potential solutions in reducing the environmental impact of air travel.

2. Support innovation: Encourage and support research and development efforts in the aviation industry that aim to find sustainable solutions. This can include advocating for government funding, engaging with industry stakeholders, and promoting environmentally friendly aviation practices.

3. Choose sustainable travel options: When planning air travel, consider airlines that prioritize fuel efficiency and sustainability. Look for carriers that actively invest in emission reduction measures and support carbon offset programs.

4. Promote awareness: Share information about emission-free flying and alternative technologies with friends, family, and colleagues. Promote awareness about the environmental impact of air travel and the need for sustainable practices within the aviation industry.

5. Reduce personal carbon footprint: While the focus is on the aviation industry, individuals can also contribute to reducing their own carbon footprint. Consider alternative modes of transportation for shorter trips, explore local travel options, and offset personal carbon emissions when flying.

6. Engage in sustainable tourism: Consider the environmental impact when planning trips. Choose eco-friendly accommodations, support local businesses, and explore sustainable tourism practices that minimize harm to the environment.

7. Support renewable energy: Advocate for the development and use of renewable energy sources, such as wind and solar power, which can contribute to the production of green hydrogen and other sustainable aviation fuels.

8. Encourage policy changes: Engage with policymakers and support initiatives that promote sustainable aviation practices. Encourage the adoption of regulations and incentives that drive the industry towards emission-free flying and investment in green technologies.

9. Foster collaboration: Recognize that achieving emission-free flying requires collaboration between governments, industry stakeholders, researchers, and the public. Support collaboration efforts and contribute to the collective goal of a greener aviation industry.

10. Be a responsible traveler: Practice responsible travel behavior by respecting local cultures, minimizing waste, and supporting local conservation efforts. Be conscious of the environmental impact of your travel choices and strive to leave a positive footprint wherever you go.

11. By incorporating these culture tips into our lives, we can contribute to a more sustainable

future for air travel and the environment.

Reading Exercises

I. Vocabulary Matching

Directions: *In this section, there is a passage with ten blanks. You are required to select one word for each blank from a list of choices given in a word bank following the passage. Read the passage through carefully before making your choices. Each choice in the blank is identified by a letter.*

More than 20,000 planes were in operation around the world before the pandemic, carrying over 4.5 billion passengers in 2019. Most of them were frequent flyers on domestic flights in the US and China. And they __1__ for about a quarter of all aviation emissions.

Flying was so exclusive, because it was very __2__. "The airplane is just like a limousine, I feel like a __3__." An ordinary U.S. domestic roundtrip cost around $600 back in the 1970s, a whopping 4,000 bucks in today's money. But then the market was deregulated: Low-cost carriers emerged, and __4__ became target No.1. The aviation industry is growing very fast and aviation evolved from the privilege of a few into a service for many.

Until 2020, when the coronavirus hit, grounding most of the world's aircrafts. But the industry is expected to bounce back soon. "We think that we will see a __5__ to pre-COVID levels of traffic by the end of 2024. But does it have to involve blasting fossil fuel emissions into the __6__? Perhaps not, hydrogen energy was __7__. Hydrogen has at least three striking features. Number one: It's a versatile energy carrier and can power aircrafts with high energy demand where batteries would be no alternative. Hydrogen can be made using __8__ energy, meaning the production can be environmentally friendly. This so-called green hydrogen is what they want. "What we are talking about here is really powering aviation with renewable energy." It does not pollute. Hydrogen when __9__ emits no CO_2 and almost no air pollution, which of course, is the biggest plus.

And finally, there's simply not enough green hydrogen available yet. Hydrogen will need more time to reach its potential. But we can help __10__ emissions now by making choices about how we travel and traveling less whenever possible. Because technology alone will not be able to save our planet.

A. discovered   B. pollution   C. renewable   D. fossil   E. millionaire

| F. expensive | G. accounted | H. commercial | I. burned | J. growth |
| K. supposed | L. reduce | M. air | N. emissions | O. return |

II. Answer the Following Questions

1. What is the current contribution of aviation emissions to global emissions?
2. What are some alternative technologies proposed to reduce emissions in the aviation industry?
3. Which airline faced legal action for misleading marketing claims about its biofuel usage?
4. What is one of the notable qualities of hydrogen as a potential alternative fuel for aviation?
5. What combination of approaches is necessary to achieve a more eco-friendly future of aviation?

III. Multiple Choice

1. What percentage of global emissions did direct CO_2 emissions from aviation account for in 2019? ( )
   A. 1%
   B. 2%
   C. 5%
   D. 10%
2. Which alternative technology has shown promise in reducing CO_2 emissions in aviation? ( )
   A. Zeppelins
   B. Blended-wing aircraft
   C. Solar power
   D. Biofuels
3. Why was the launch of the fully electric-powered aircraft, the E-Fan 1, canceled? ( )
   A. Lack of consumer demand
   B. Technological limitations
   C. Environmental concerns
   D. Safety issues
4. What is one notable quality of hydrogen as an energy carrier for aircraft? ( )
   A. It emits greenhouse gases when used in fuel cells.
   B. It relies heavily on fossil fuels for production.
   C. It can power high-demand aircraft.

D. It has limited availability.

5. What is necessary to achieve a more eco-friendly future of aviation? (   )

   A. Improving fuel efficiency and optimizing flight routes

   B. Transitioning to biofuels and hydrogen

   C. Technological advancements and supportive infrastructure

   D. All of the above

Translation

Directions: *Translate the following sentences into Chinese.*

1. Aviation has become a major contributor to the climate crisis, with its carbon footprint posing a significant threat.

   _____
   _____

2. In 2019, direct CO_2 emissions from aviation accounted for at least 2% of global emissions, making it the most energy-intensive human activity.

   _____
   _____

3. However, with the deregulation of the market, low-cost carriers emerged, catering to a growing demand for air travel. Aviation transformed from a privilege for a few into a service for many.

   _____
   _____

4. Thankfully, the aviation industry has proposed numerous alternative technologies over the years, fueling hope for a greener future.

   _____
   _____

5. Biofuels, in particular, have shown promise in reducing CO_2 emissions, but their limited availability remains a challenge.

   _____
   _____

6. Hydrogen possesses three notable qualities: versatility as an energy carrier that can power high-demand aircraft, the potential for environmentally friendly production using

Unit 11 Aviation Energy

renewable energy (green hydrogen), and the emission of only water vapor when used in fuel cells.

7. While hydrogen holds promise, it is not a one-size-fits-all solution at present.

8. Ultimately, a combination of approaches, including technological innovations and changes in consumer behavior, will be necessary to achieve a more eco-friendly future of aviation.

Speaking

Student: I've heard about hydrogen-powered planes as a solution for emission-free flying. Can you tell me more?

Aviation expert: Sure! Hydrogen-powered planes are being explored as a promising option. Hydrogen has qualities that make it suitable for aircraft, like powering high-demand planes and emitting only water vapor when used in fuel cells.

Student: What challenges are associated with hydrogen-powered planes?

Aviation expert: There are challenges. Currently, hydrogen production relies on fossil fuels, which undermines its environmental benefits. We need to scale up the production of green hydrogen using renewable sources. Plus, we're still developing the technology for hydrogen fuel cells, and we need infrastructure for hydrogen refueling.

Student: Are there other alternatives that have been considered?

Aviation expert: Yes, alternatives like zeppelins, blended-wing aircraft, solar power, and biofuels have been explored. Biofuels show promise but face limited availability. Electric aircraft also generated hype but have technical limitations.

Student: What steps is the aviation industry taking to reduce its carbon footprint for now?

Aviation expert: Airlines are improving fuel efficiency, optimizing flight routes, and offsetting emissions through carbon offset programs. These are important interim measures while we work on long-term solutions.

Student: What do you think the future holds for emission-free flying?

Aviation expert: It will require a combination of approaches. We need investments, technological advancements, and infrastructure for hydrogen-powered planes. Consumer behavior changes are also important. With ongoing research and awareness, we can achieve a greener future for aviation.

Student: That's great to hear. I hope we can make air travel more sustainable soon.

Aviation Expert: Absolutely! By working together and pursuing innovative solutions, we can move towards a more eco-friendly future of aviation. It's an exciting time, and we're on the right path.

Writing

Directions: *Write a composition entitled "Save the Energy Resources". You should write at least 120 words according to the outline given below in Chinese.*

1. 现在很多能源资源面临枯竭；
2. 阐述能源资源枯竭的后果；
3. 节约能源资源的建议。

_____
_____
_____
_____
_____
_____
_____

Expansion Exercise for CET-BAND 4—News Report

Glasgow has pledged to become the first carbon neutral city in the UK. The city's council and ScottishPower have announced a range of strategies in an attempt to reduce carbon emissions ahead of the new national target of 2045.

First Minister Nicola Sturgeon welcomed the pledge and said: "Today's announcement between ScottishPower and Glasgow City Council—to make Glasgow the UK's first net-zero city—is a very welcome step. Reaching our goals will need exactly this kind of partnership approach—with government, business, local authorities and citizens all playing their part."

Speaking ahead of the All Energy Conference being held in Glasgow, ScottishPower chief executive Keith Anderson said: "We have a large supply of renewable energy on our doorstep and one of the only two low emission zones in action across the UK. Now, we need to invest in the technologies and programmes that transform the rest of Glasgow's economy and make us net zero before anyone else."

1. What do we learn from the news report? (　)
    A. Scotland will reach the national target in carbon emissions reduction ahead of schedule.
    B. Glasgow City Council has made a deal with ScottishPower on carbon emissions.
    C. Glasgow has pledged to take the lead in reducing carbon emissions in the UK.
    D. First Minister Nicola Sturgeon urged ScottishPower to reduce carbon emissions.
2. What did ScottishPower's chief executive say ahead of the All Energy Conference?
　　　　　　　　　　　　　　　　　　　　　　　　　　　　　　　　　　　　(　)
    A. Glasgow needs to invest in new technologies to reach its goal.
    B. Glasgow is going to explore new sources of renewable energy.
    C. Stricter regulation is needed in transforming Glasgow's economy.
    D. It's necessary to create more low-emission zones as soon as possible.

Unit 12 Aviation Accident

Preview

On September 7, 2010, Flight 516 experienced a critical electrical failure while flying from Udachny, Russia, to Moscow. With all electronics and navigational systems disabled, the pilots resorted to using a glass of water to measure the angle of descent. They spotted an abandoned runway in Izhma, thanks to the efforts of airport supervisor Sergey Sotnikov, and executed a challenging landing. Despite the landing gear almost catching fire and the plane crashing into the forest, all 72 passengers and 9 crew members survived without injuries. The heroic pilots and cabin crew were honored, and the plane continued flying for another 7 years before retiring in 2018.

Listening

Directions: *Fill in the blanks with the words that you hear. Find them in the table below and write down the corresponding letters.*

Flight 516, scheduled from Udachny, Russia, to Moscow, encountered a life-threatening situation in 2010. Although the plane had operated without incidents for 20 years, it faced unexpected __1__ failures mid-flight. The pilots bravely fought against the odds to land the aircraft safely, saving the lives of all 72 passengers and crew members on board.

As the plane passed over Usinsk in Siberia, the electronic equipment started malfunctioning. Initially dismissing it as a minor inconvenience, the pilots continued their journey. However, the situation quickly escalated when all the electronics on board shut down, leaving the pilots without crucial navigational systems, communication, and control. With limited fuel and visibility, they had to rely on their instincts and experience to guide the plane.

Descending through the clouds, the pilots faced a desperate situation. The lack of functioning instruments made finding a suitable landing spot nearly impossible. To make matters worse, the runway they eventually spotted near Izhma was abandoned and too short for a safe landing. The plane's speed exceeded the __2__ limits, posing a significant risk.

In a stroke of ingenuity, the pilots used a glass of water as an improvised inclinometer to gauge their descent. This resourceful solution allowed them to navigate closer to the ground, enabling them to activate the __3__ generator. However, their attempts to restore power failed, leaving them with limited options.

Miraculously, they spotted the abandoned runway near Izhma, meticulously maintained by airport __4__ Sergey Sotnikov. Against all odds, the plane touched down on the runway, albeit with the landing gear ablaze due to the friction. The aircraft finally came to a halt after crashing into the forest, sustaining damage but miraculously sparing everyone on board from harm.

The heroic actions of the pilots and the calmness displayed by the passengers and crew during this ordeal prevented a __5__. Following the incident, decorations and honors were awarded to recognize their professionalism and bravery. Sergey Novosyolov and Andrey Lamanov, the pilots, were honored as Heroes of the Russian Federation, while the rest of the cabin crew __6__ the orders of courage. Sergey Sotnikov, the airport supervisor, was also recognized for his dedication.

Unit 12 Aviation Accident

After the incident, there were doubts about the future of the damaged plane. Initially considered for dismantling due to its __7__ to take off from the short runway in Izhma, a decision was eventually made to rescue and repair it. Following repairs, the aircraft continued to serve for seven more years, earning the nickname "Izhma". In September 2018, after nearly three __8__ of service, the plane retired.

The remarkable story of Flight 516 serves as a testament to the bravery, __9__, and professionalism of those involved. It stands as a reminder that even in the face of adversity, human __10__ and ingenuity can triumph over seemingly insurmountable challenges.

A. received   B. determination   C. decades   D. catastrophe   E. electronic
F. recommended   G. auxiliary   H. inability   I. supervisor   J. resourcefulness

Reading

A Glass of Water Saved a Plane and 72 Passengers

On September 7, 2010, Flight 516 **embarked** on its scheduled journey from Udachny, Russia, to Moscow. The aircraft, having served for two decades, **boasted** an **impeccable** safety record with no previous accidents. The cabin crew **exuded** confidence, as they had no reason to **doubt** that this flight would be any different.

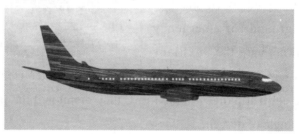

Passengers boarded the plane, and the flight attendants casually checked their boarding passes, following routine procedures. The first pilot warmly greeted everyone, signaling the beginning of the flight. Everything seemed ordinary until the aircraft reached a small Siberian town called Usinsk.

It was over Usinsk that the crew first noticed signs of trouble. The electronic equipment on board began behaving **erratically**, causing some concern. However, both the pilots and air traffic control initially **dismissed** it as a mere **nuisance**. Captain Yevgeny Novosyolov assured the passengers that they were experiencing slight turbulence and that the flight might become a bit rough temporarily. Remarkably, his **reassurance** worked, and the passengers remained calm and trusting of their skilled pilot.

However, just eight minutes after Captain Novosyolov's **announcement**, the lights in the cabin suddenly went out. While the flight attendants didn't pay much attention, believing it to be a minor issue, the cockpit was **engulfed** in **desperation**. It wasn't just the lights that failed; all the **electronics on board** suddenly shut down, leaving the pilots completely reliant on their own expertise. The plane remained operational, but without **functioning navigational system**s, **radio communication**, or **fuel transfer pumps**.

This **critical** situation left the pilots with the **daunting** task of navigating using only their visual cues. **Compounding** the challenge, the remaining usable fuel could only sustain the aircraft in the air for approximately 30 minutes. Time **was of the essence**, and it became evident that reaching the nearest airport, even under ideal circumstances, would be impossible. With all electronic systems **rendered** useless, finding a suitable landing spot appeared to be an **insurmountable** feat.

Undeterred by the **dire** circumstances, the pilots drew upon their experience and skill to maintain the stability of the plane. At the time of the electrical failure, the aircraft found itself soaring high above the clouds at an altitude of 34,780 feet (10,600 meters). To **descend** below the cloud cover, enabling them to assess the ground for emergency landing options and activate the **auxiliary generator**, the pilots resorted to an **ingenious** yet almost forgotten technique: they placed a glass of water on the **control panel**. Taking turns, they monitored the **incline** of the water's surface to **gauge** the angle of descent.

This **resourceful** solution allowed them to descend to 9,840 feet (3,000 meters), finally reaching a level where they could attempt to activate the generator. However, their efforts proved **futile**, as the electrical failure **persisted**. Nevertheless, the crew refused to **succumb** to despair and continued searching for a suitable landing site. It was during this critical moment that they spotted an abandoned **landing strip** near the town of Izhma.

The discovery of the landing strip was **nothing short of** a **miracle**. Unmarked on current maps, this old airport had long been out of operation, except for **helicopter** landings. Yet, airport **supervisor** Sergey Sotnikov, driven by his own **dedication**, had **diligently** maintained the runway's condition over the years, clearing overgrown vegetation and ensuring its usability. This unforeseen **stroke of luck** allowed the pilots to consider the possibility of a safe landing.

However, there were significant challenges to overcome. The length of the Izhma runway was only 4,347 feet (1,325 meters), **falling short of** the required 6,560 feet (2,000 meters) needed for a safe landing of such a large and heavy aircraft. Additionally, the airplane's speed during descent far exceeded the recommended limits, with a **velocity** of 236 mph (380 kph) compared to the safe **threshold** of 168 mph (270 kph). The **flaps**, crucial for controlling the aircraft's descent, could only be operated via electrical switches.

Despite the **odds stacked against** them, the cabin crew now had a **glimmer** of hope. No longer faced with the prospect of a forest landing that could **jeopardize** everyone's lives, they seized the unexpected opportunity. The pilots made three **determined** attempts to land the aircraft safely, with the first two **necessitating** a return to the air for another circuit. Their cautious approach, evaluating the risks **meticulously**, undoubtedly played a **pivotal** role in ensuring the passengers' safety.

Before the final landing attempt, all passengers and the cabin crew were relocated to the front of the aircraft, preparing for a potential **emergency evacuation**. Remarkably, there was no panic among the passengers. Everyone remained calm and **composed**, fully aware of the gravity of the situation and the **imminent** danger they faced. They patiently awaited their fate, maintaining silence throughout the **ordeal**.

Finally, the moment of truth arrived. The pilots committed to the landing, bringing the aircraft down onto the runway. The friction generated by the landing nearly **ignited** the landing gear, yet it failed to halt the plane before the runway ended. The pilots skillfully applied maximum **reverse thrust** and engaged the emergency brakes, desperately attempting to decelerate the aircraft as quickly as possible. The plane **hurtled** forward, inching closer to the end of the runway.

With every second that passed, the tension inside the cabin grew **palpable**. The passengers braced themselves, holding onto their seats tightly, praying for a miracle. The **screeching** sound of tires against the **asphalt reverberated** throughout the aircraft as the plane raced towards the edge of the runway.

In a final act of **sheer determination**, the pilots managed to bring the aircraft to a complete stop just a few meters from the end of the runway. The cabin erupted in a mixture of relief, disbelief, and **jubilation**. The passengers broke into applause, applauding the incredible skill and nerve of the pilots who had **defied** the odds and saved their lives.

Emergency services quickly arrived at the scene to evacuate the passengers and crew.

Despite the **intensity** of the landing, there were no serious injuries reported. The passengers, filled with a **renewed** appreciation for life, hugged their loved ones and expressed their **gratitude** to the pilots and crew who had safely guided them through this **harrowing** experience.

The crew's **professionalism** and quick thinking were **commended** by aviation experts and authorities. Captain Yevgeny Novosyolov and his co-pilot, Dmitry Rodin, were **hailed** as heroes for their exceptional skills and the successful outcome of the flight. Their ability to adapt and remain composed in the face of extreme adversity had undoubtedly played a crucial role in the survival of all 72 passengers and 9 crew members on board.

Following the incident, a thorough **investigation** was conducted to determine the cause of the electrical failure and subsequent equipment malfunctions. It was traced back to a **faulty** component in the electrical system, which had gradually **deteriorated** over time. As a result, the airline **implemented comprehensive** maintenance and inspection protocols to prevent similar incidents in the future.

The aircraft involved in Flight 516, a Tupolev Tu-154, underwent extensive repairs and returned to service after several weeks. It continued flying for another seven years, ensuring the safety of its passengers until its retirement in 2018.

The successful landing of Flight 516 near Izhma remains a remarkable **testament** to the skill, resourcefulness, and professionalism of the pilots and crew. Their **unwavering** determination to save lives in the face of **overwhelming** odds serves as an **inspiration** to aviation professionals and a reminder of the importance of human expertise in the age of advanced technology.

New Words and Expressions

embark *v.* /ɪmˈbɑːk/	to start to do sth. new or difficult 从事，着手，开始
boast *v.* /bəʊst/	to have sth. that is impressive and that you can be proud of 有（值得自豪的东西）
impeccable *adj.* /ɪmˈpekəbl/	without mistakes or faults 无错误的，无瑕疵的，完美的
exude *v.* /ɪgˈzjuːd/	show a particular feeling or quality 流露，显露，显现
doubt *n.* /daʊt/	feelings of uncertainty or lack of conviction

怀疑

erratically *adv.* /ɪˈrætɪkli/	variably, abnormally, in a way that deviates from the usual or typical
	异常地
dismiss *v.* /dɪsˈmɪs/	to decide that sb./sth. is not important and not worth thinking or talking about
	不予考虑，摒弃，对……不屑
nuisance *n.* /ˈnjuːsns/	a thing, person or situation that is annoying or causes trouble or problems
	麻烦事，讨厌的人（或东西）
reassurance *n.* /ˌriːəˈʃʊərəns/	the fact of giving advice or help that takes away a person's fears or doubts
	（能消除疑虑等的）肯定，保证
announcement *n.* /əˈnaʊnsmənt/	a formal public statement about a fact, occurrence, or intention.
	宣布
engulf *v.* /ɪnˈɡʌlf/	to surround or to cover sb./sth. completely
	包围，吞没，淹没
desperation *n.* /ˌdespəˈreɪʃn/	the state of being desperate
	绝望，拼命，铤而走险
function *v.* /ˈfʌŋkʃ(ə)n/	to work in the correct way
	起作用，正常工作，运转
critical *adj.* /ˈkrɪtɪkəl/	expressing adverse or disapproving comments or judgments
	关键的
daunting *adj.* /ˈdɔːntɪŋ/	seeming difficult to deal with in prospect; intimidating
	使人畏惧的，使人气馁的
compound *v.* /kəmˈpaʊnd/	to make sth. bad become even worse by causing further damage or problems
	使加重，使恶化
render *v.* /ˈrendə(r)/	to cause sb./sth. to be in a particular state or condition
	使成为，使变得，使处于某状态

insurmountable *adj.* /ˌɪnsəˈmaʊntəbl/	that cannot be dealt with successfully
	无法克服的，难以解决的，不可逾越的
undeterred *adj.* /ˌʌndɪˈtɜːd/	have strong will in doing sth. and will not be stopped
	顽强的，坚毅的，不屈不挠的
dire *adj.* /ˈdaɪə(r)/	very serious, very bad
	极其严重的，危急的，极糟的，极差的
descend *v.* /dɪˈsend/	moved or passed from a higher to a lower place or level
	下降
ingenious *adj.* /ɪnˈdʒiːniəs/	very suitable for a particular purpose and resulting from clever new ideas
	精巧的，新颖独特的，巧妙的
incline *n.* /ˈɪnklaɪn/	a slope
	斜坡，倾斜，斜度
gauge *v.* /ɡeɪdʒ/	to calculate sth. approximately
	估计，估算
resourceful *adj.* /rɪˈsɔːsfl/	good at finding ways of doing things and solving problems, etc.
	机敏的，足智多谋的，随机应变的
futile *adj.* /ˈfjuːtaɪl/	having no purpose because there is no chance of success
	徒然的，徒劳的，无效的
persist *v.* /pəˈsɪst/	to continue to exist
	维持，保持，持续存在
succumb *v.* /səˈkʌm/	to not be able to fight an attack, an illness, a temptation, etc.
	屈服，屈从，抵挡不住（攻击、疾病、诱惑等）
miracle *n.* /ˈmɪrəkəl/	a remarkable event or phenomenon that cannot be explained by natural or scientific laws and is therefore attributed to a divine agency.
	奇迹
helicopter *n.* /ˈhelɪˌkɒptə/	a type of aircraft that derives its lift from

	rotating blades
	直升机
supervisor *n.* /ˈsuːpəvaɪzə/	a person who supervises sb./sth.
	监督人，指导者，主管人
dedication *n.* /ˌdedɪˈkeɪʃn/	the hard work and effort that sb. puts into an activity or purpose because they think it is important
	献身，奉献
diligently *adv.* /ˈdɪlɪdʒəntli/	with hard work
	勤奋地，勤勉地
velocity *n.* /vəˈlɒsəti/	the speed of sth. in a particular direction
	（沿某一方向的）速度
threshold *n.* /ˈθreʃhəʊld/	the level at which sth. starts to happen or have an effect
	阈，界，起始点
flap *n.* /flæp/	a part of the wing of an aircraft that can be moved up or down to control upward or downward movement
	（飞行器的）襟翼
odds *n.* /ɒdz/	希望，可能性
glimmer *n.* /ˈglɪmə(r)/	a small sign of sth.
	微弱的迹象，一丝，一线
jeopardize *v.* /ˈdʒepədaɪz/	to risk harming or destroying sth./sb.
	冒……的危险，危及，危害，损害
determined *adj.* /dɪˈtɜːmɪnd/	showing a person's will to do sth.
	决意的，下决心的
necessitate *v.* /nəˈsesɪteɪt/	to make sth. necessary
	使成为必要
meticulously *adv.* /məˈtɪkjələsli/	very carefully
	细致地，一丝不苟地，拘泥地
pivotal *adj.* /ˈpɪvətl/	of great importance because other things depend on it
	关键性的，核心的
composed *adj.* /kəmˈpəʊzd/	calm and in control of your feelings

	镇静，镇定，平静
imminent *adj.* /ˈɪmɪnənt/	likely to happen very soon
	即将发生的，临近的
ordeal *n.* /ˈɔːdiːl/	a difficult or unpleasant experience
	磨难，折磨，煎熬，严酷的考验
ignite *v.* /ɪɡˈnaɪt/	to start to burn; to make sth. start to burn
	（使）燃烧，着火，点燃
investigation *n.* /ɪnˌvestəˈɡeɪʃən/	the action of looking into sth., typically to uncover facts or information.
	调查
hurtle *v.* /ˈhɜːtl/	to move very fast in a particular direction
	（向某个方向）飞驰，猛冲
palpable *adj.* /ˈpælpəbl/	that is easily noticed by the mind or senses
	易于察觉的，可意识到的，明显的
screech *v.* /skriːtʃ/	to make a loud high unpleasant sound; to say sth. using this sound
	尖叫，发出尖锐刺耳的声音，尖声地说
asphalt *n.* /ˈæsfælt/	a thick black sticky substance used especially for making the surface of roads
	沥青，柏油
reverberate *v.* /rɪˈvɜːbəreɪt/	to be repeated several times as it is reflected off different surfaces
	回响，回荡
sheer *adj.* /ʃɪə(r)/	complete and not mixed with anything else
	完全的，纯粹的，十足的
determination *n.* /dɪˌtɜːmɪˈneɪʃn/	the quality that makes you continue trying to do sth. even when this is difficult
	决心，果断，坚定
jubilation *n.* /ˌdʒuːbɪˈleɪʃn/	a feeling of great happiness because of a success
	欢欣鼓舞，欢腾，欢庆
defy *v.* /dɪˈfaɪ/	to refuse to obey or show respect for sb. in authority, a law, a rule, etc.
	违抗，反抗，蔑视

intensity *n.* /ɪnˈtensəti/	the state or quality of being intense
	强烈，紧张，剧烈
renewed *adj.* /rɪˈnjuːd/	happening again with increased interest or strength
	再次发生的，再次兴起的，更新的
gratitude *n.* /ˈɡrætɪtjuːd/	the feeling of being grateful and wanting to express your thanks
	感激之情，感谢
harrowing *adj.* /ˈhærəʊɪŋ/	very shocking or frightening and making you feel very upset
	使人十分难过的，恐怖的
professionalism *n.* /prəˈfeʃənəˌlɪzəm/	the competence or skill expected of a professional.
	专业素养
commend *v.* /kəˈmend/	to praise sb./sth., especially publicly
	（尤指公开地）赞扬，称赞，表扬
hail *v.* /heɪl/	to describe sb./sth. as being very good or special, especially in newspapers, etc.
	赞扬（或称颂）……为……（尤用于报章等）
faulty *adj.* /ˈfɔːlti/	not perfect; not working or made correctly
	不完美的，有错误的，有缺陷的
deteriorate *v.* /dɪˈtɪəriəreɪt/	to become worse
	变坏，恶化，退化
implement *v.* /ˈɪmplɪment/	to make sth. that has been officially decided start to happen or be used
	使生效，贯彻，执行，实施
comprehensive *adj.* /ˌkɒmprɪˈhensɪv/	including all, or almost all, the items, details, facts, information, etc., that may be concerned
	全部的，所有的，（几乎）无所不包的，详尽的
testament *n.* /ˈtestəmənt/	a thing that shows that sth. else exists or is true
	证据，证明
unwavering *adj.* /ʌnˈweɪvərɪŋ/	not changing or becoming weaker in any way

	不动摇的，坚定的，始终如一的
overwhelming *adj.* /ˌəʊvəˈwelmɪŋ/	very great or very strong; so powerful that you cannot resist it or decide how to react
	巨大的，压倒性的，无法抗拒的
inspiration *n.* /ˌɪnspəˈreɪʃn/	sb. or sth. that makes you want to be better, more successful, etc.
	鼓舞人心的人（或事物）
electronics on board	机载电子设备
navigational systems	导航系统
radio communication	无线电通信
fuel transfer pump	燃油输送泵
visual cue	视觉提示
be of the essence	至关重要的事情
emergency landing	紧急着陆
auxiliary generator	辅助发电机
control panel	控制面板
landing strip	着陆跑道
nothing short of	简直就是，无异于
stroke of luck	幸运的一击
fall short of	未达到预期的标准、目标或期望
stack against	操纵某事使其不利于某人
emergency evacuation	紧急疏散
safety instructions	安全指示
reverse thrust	反推力
emergency service	紧急服务

Genre Analysis

Genre:
A narrative or a recounting of a real-life event. It tells the story of Flight 516 and its emergency landing, highlighting the challenges faced by the pilots and crew and the successful outcome of the flight.

Communicative Moves:
Move 1 Setting the Scene:
The passage begins by providing background information about Flight 516, its safety record,

and the confidence of the cabin crew. It sets the stage for the events that follow.

Move 2 Introducing the Problem:

The passage gradually introduces the signs of trouble on the aircraft, highlighting the sudden electrical failure and the critical situation faced by the pilots.

Move 3 Describing the Solution:

The passage details the resourcefulness of the pilots in using a glass of water to gauge the angle of descent and their discovery of an abandoned landing strip as a potential safe landing site.

Move 4 Outlining Challenges:

The passage outlines the challenges faced by the pilots, such as the short runway length and excessive speed during descent.

Move 5 Building Suspense:

The passage builds tension as the plane approaches the end of the runway, with passengers bracing themselves and praying for a safe outcome.

Move 6 Resolution and Conclusion:

The passage describes the successful landing, the relief and jubilation of the passengers, and the commendation of the crew. It also highlights the subsequent investigation and implementation of safety measures.

Communicative Purpose:

The purpose of the passage is to recount a real-life incident involving Flight 516, emphasizing the skill, resourcefulness, and professionalism of the pilots and crew. It aims to engage and captivate readers with a compelling narrative of overcoming adversity in a high-stakes situation. Additionally, it highlights the importance of human expertise and decision-making in the face of technological failures and serves as a testament to the dedication and courage of aviation professionals.

Culture Tips

1. Respect and trust professionals: The passage highlights the professionalism and expertise of the pilots and cabin crew. In various fields, it's important to respect and trust professionals who have undergone extensive training and have experience in their respective domains.

2. Admire resourcefulness: The pilots in the story displayed resourcefulness by using a glass of water to aid their descent. This reminds us of the value of ingenuity and thinking outside the box when faced with challenging situations. Cultivate a mindset of creativity and adaptability in problem-solving.

3. Remain calm in emergencies: The passengers in Flight 516 remained remarkably calm throughout the ordeal. This underscores the importance of maintaining composure during crises. Cultivate resilience and emotional stability in the face of unexpected challenges.

4. Appreciate human expertise: Despite technological failures, the pilots relied on their own skills and experience to navigate and land the aircraft safely. This emphasizes the importance of human expertise and decision-making in critical situations. Acknowledge the value of human judgment and expertise alongside technological advancements.

5. Support and encourage teamwork: The successful outcome of Flight 516 was a result of effective teamwork between the pilots and cabin crew. Encourage and appreciate teamwork in various settings, recognizing the strength and synergy that can emerge when individuals work together towards a common goal.

6. Embrace safety measures: The incident led to subsequent investigations and the implementation of safety measures. This highlights the importance of adhering to safety protocols and continuously improving safety standards in various industries. Prioritize safety in your personal and professional endeavors.

7. Acknowledge and learn from extraordinary stories: Extraordinary stories like Flight 516 serve as reminders of the resilience, determination, and courage of individuals facing adversity. Take the opportunity to learn from these stories, draw inspiration from them, and apply the lessons learned to your own life.

8. Remember, culture tips can vary based on specific contexts and perspectives. However, these general tips can help foster a culture of professionalism, resourcefulness, calmness, and safety awareness.

Reading Exercises:

I. Vocabulary Matching

Directions: *Fill in the blank with the corresponding letter of the Chinese equivalent of the following English words.*

1. travel   (   )   7. interpret   (   )   13. necessary   (   )   19. traffic   (   )
2. abandonment (   )   8. estimate   (   )   14. neglect   (   )   20. trouble   (   )
3. absolutely   (   )   9. departure   (   )   15. occur   (   )   21. unable   (   )
4. advance   (   )   10. available (   )   16. official   (   )   22. understanding (   )
5. advantage   (   )   11. status   (   )   17. opportunity (   )
6. against   (   )   12. procedure (   )   18. perfect   (   )

A. 离开	G. 交通	M. 不能	S. 解释
B. 放弃	H. 估计	N. 被忽视的	T. 推进
C. 绝对	I. 旅行	O. 发生	U. 必要的
D. 麻烦	J. 可用	P. 机会	V. 理解
E. 优势	K. 状态	Q. 官方的	
F. 反对	L. 过程	R. 完美的	

II. Answer the Following Questions

1. How did the pilots navigate the plane after the electronic equipment malfunctioned?
2. What challenges did the pilots face during the landing attempt?
3. How did the pilots manage to land the aircraft safely?
4. How did the passengers react during the ordeal?
5. What was the cause of the equipment malfunction on Flight 516?

III. Multiple Choice

1. What technique did the pilots use to assess the angle of descent during the electrical failure? （　）

    A. They relied on their visual cues.

    B. They used a radar system.

    C. They communicated with air traffic control.

    D. They placed a glass of water on the control panel.

2. How did the pilots discover the abandoned landing strip near the town of Izhma? （　）

    A. They noticed it on their navigational systems.

    B. The air traffic control directed them to it.

    C. They saw it from the cockpit window.

    D. Passengers alerted them to it.

3. What were the challenges the pilots faced when attempting to land on the Izhma runway?

    （　）

    A. The runway was too short, and the aircraft's speed was too high.

    B. The runway was covered in overgrown vegetation.

    C. The landing gear was malfunctioning.

    D. The pilots couldn't communicate with the control tower.

4. How did the passengers react during the landing attempts? （　）

    A. They panicked and screamed.

    B. They remained calm and composed.

C. They protested and demanded answers.

D. They attempted to exit the aircraft prematurely.

5. What was the outcome of the landing of Flight 516 near Izhma? ( )

   A. The plane crashed, resulting in numerous injuries.

   B. The plane came to a stop just a few meters from the end of the runway.

   C. The passengers had to be rescued from the forest.

   D. The landing gear caught fire, but everyone was safely evacuated.

Translation

Directions: *Translate the following sentences.*

1. Passengers got on board the plane; flight attendants casually checked their boarding passes; everything went as usual.

2. Captain Yevgeny Novosyolov assured the passengers that the plane was going through a slight turbulence, and the flight could get a little rough for a short time.

3. It was still operable, but all the navigational systems refused to work, the radio went offline, and the fuel transfer pumps were also lost.

4. The time was not enough for the flight to get to the nearest airport even in otherwise perfect condition, but with all the electronics gone, finding a good spot to land—and land there safely, for that matter—was next to impossible.

5. 飞行员们轮流观察，看着玻璃杯里的水，通过其倾斜度来测量下降角度。

6. 尽管机场被弃置，但索特尼科夫仍凭借自己的意愿，让跑道保持了多年的良好状态。

Unit 12 Aviation Accident

7. 然而，伊热尔曼机场的跑道长度只有 4347 英尺（1325 米），而这架飞机需要 6560 英尺（2000 米）的跑道来安全着陆。

8. 因此，在所有不利情况下，尽管飞机的电气系统都不工作、没有适当的跑道可以降落、速度快到原本很可能撞进森林，甚至没有办法向其他任何人预警，英勇的飞行员仍然成功降落了飞机，挽救了每个人的生命。

9. 面对巨大的困难，他们拯救生命的坚定决心鼓舞了其他航空专业人士，并提醒人们即使处在先进技术的时代，人类的专业知识仍然不可替代。

Speaking

Student: Excuse me, Captain Smith, I recently read an incredible story about a flight where the pilots had to rely on their own expertise and resourcefulness to save the passengers. It was Flight 516. Have you heard about it?

Pilot: Yes, I'm familiar with that incident. It was quite a remarkable story. The pilots faced a challenging situation when all the electronics on board suddenly shut down, leaving them without navigational systems or radio communication.

Student: That must have been terrifying. How did they manage to land the plane safely without any functioning equipment?

Pilot: They had to rely on their visual cues and their own experience to navigate. But what's truly ingenious is how they used a simple glass of water on the control panel. By monitoring the incline of the water's surface, they were able to gauge the angle of descent and descend below the cloud cover.

Student: That's incredible! It's amazing how they found a solution with such limited resources. Did they face any other challenges during the landing?

Pilot: Absolutely. The pilots discovered an abandoned landing strip near a town called Izhma, but it fell short of the required length for a safe landing. Additionally, the plane's speed during descent exceeded the recommended limits. The pilots had to carefully maneuver the aircraft while dealing with these limitations.

Student: It must have been a nerve-wracking experience for the passengers. How did they handle the situation?

Pilot: Surprisingly, the passengers remained remarkably calm and composed throughout the ordeal. They were aware of the gravity of the situation and patiently awaited their fate. They were relocated to the front of the aircraft, preparing for a potential emergency evacuation.

Student: It's impressive how everyone remained so calm. And what happened during the actual landing?

Pilot: Well, the pilots made three attempts to land the aircraft safely. It was a tense moment as the plane approached the end of the runway, but they managed to bring it to a complete stop just a few meters away. The passengers erupted in relief and applause, grateful for the skill and determination of the pilots.

Student: That's truly amazing! It's a testament to the pilots' expertise and professionalism. I can't imagine the relief everyone must have felt when they realized they were safe.

Pilot: Indeed, it was a challenging situation, but the pilots' quick thinking and their ability to adapt played a crucial role in the successful outcome. It's a reminder of the importance of human skills and experience in the face of unexpected emergencies.

Student: Absolutely! This story has given me a newfound appreciation for the work and dedication of pilots like yourself. Thank you for sharing your insights, Captain Smith.

Pilot: You're welcome, and remember, safety and preparedness are always paramount in aviation. If you have any more questions, feel free to ask.

Writing

Directions: *Assume that you witnessed a traffic accident. Write a composition entitled "An Eye-witness Account of a Traffic Accident". You should write at least 120 words according to the outline given below in Chinese.*

1. 车祸发生的时间及地点；
2. 你所见到的车祸情况；
3. 你对车祸原因的分析。

Unit 12 Aviation Accident

Expansion Exercise for CET-BAND 4—News Report

  Drivers on their way to the Polish capital of Warsaw on Wednesday morning found the road blocked by an unusual obstacle: tons of liquid chocolate that spilled onto the motorway. A truck carrying the sweet load hit a road barrier and overturned, blocking two lanes. The cracked tank spilled a pool of rapidly-hardening chocolate, which quickly covered the width of the road. While the driver was taken to hospital with a broken arm, firefighters struggled to remove a reported twelve tons of solid chocolate from the road. A representative for the firefighters told the local TV that removing the chocolate was worse than dealing with snow. After contacting the chocolate manufacturer, the firefighters resorted to spraying hot, pressurized water to get rid of the sticky substance. The local TV also noted that the cleanup spanned more than a mile, because drivers simply drove through the chocolate after the crash, leaving a long chocolate trail. But despite the sticky situation, firefighters and police attending to the cleanup were reportedly cheerful about the long task ahead. After all, who could be mad about twelve tons of chocolate?

1. What did drivers on the motorway to Warsaw find? (　)
   A. The road was flooded.
   B. The road was blocked.
   C. The road was frozen with snow.
   D. The road was covered with spilled gas.
2. What does the report say about the accident? (　)
   A. A truck plunged into a pool of liquid chocolate.
   B. The heavy snow made driving very difficult.
   C. The truck driver dozed off while driving.
   D. A truck hit a barrier and overturned.

3. What did the firefighters' representative tell the local TV? (　)

   A. It was a long time before the cleanup was finished.

   B. It was a hard task to remove the spilled substance.

   C. It was fortunate that no passenger got injured.

   D. It was difficult to contact the manufacturer.

Keys

Unit 1

Listening

1~5 G B F J D

6~10 I A E H C

Reading Exercises

I. Vocabulary Matching

1~5   W U Q R S

6~10   O Y H I J

11~15 M V P X D

16~20 E F G K L

21~25 T B C A N

II. Answer the Following Questions

1 AOK-pass is a digital health passport app. It verifies COVID-19 test results or vaccination status. Travelers input their information and get a unique QR code as proof. The code is scanned during check-in and boarding.

2 The travel industry wants to recover from the pandemic's impact. Airports and airlines are exploring digital health passports to compete in the $20 billion travel health services market.

3 AOK-pass uses blockchain technology, making each test result and vaccination certificate unique and tamper-proof.

4 Data privacy and leaks are concerns. AOK-pass encrypts data and collects minimal information, but measures are needed to protect privacy and avoid stigma.

5 AOK-pass will launch in major Italian and French airports. It faces competition from apps like Common Pass and IBM's digital health pass but has signed contracts with airports and airlines for testing and development.

III. Multiple Choice

1~5   A A C C A

Translation

1 旅游业专家表示，这种类型的数字健康证书可能会引起变革，不仅对旅游业，对现场活动和学校也是如此。

2 在新冠疫情背景下，健康信息的使用起到的转型过渡作用比以往更加明显。

3 我向你展示我的部分医疗健康信息，以便登机。

4 政策制定者应要求数字健康卫生通行码提供者采取适当的技术和编制方法，切实解决潜在的数据隐私侵犯问题。

5 要把以下过程协调一致非常不易：首先在某国的检测机构创建一个健康档案；然后登上飞机（航空业也是高度管制的产业）；接着入境另一国，该国施行的法律法规与你通过健康检测的国家并不相同。

Expansion Exercise for CET-BAND 4—Reading

1~5　D A B C D

Unit 2

Listening

1~5　B I G E A

6~10　C D F J H

Reading Exercises

I. Vocabulary Matching

1~5　K H B D E

6~10　I T W C J

11~15　Q O N P M

16~20　U X A S V

21~24　F R L G

II. Answer the Following Questions

1 The four forces are weight, lift, drag, and thrust.

2 Aerodynamic lift enables the aircraft to go to altitude by countering the downward force of weight.

3 The speed of sound at sea level is approximately 760 miles per hour.

4 A sonic boom is caused by the instantaneous change in pressure as an airplane travels faster than the speed of sound, pushing air molecules out of the way and creating a high-pressure shockwave.

5 Scientists use wind tunnels to test new supersonic plane designs. Models of the designs are placed in the wind tunnel, and the pressures and data points are measured as the air is blown past the model.

III. Multiple Choice

1~5　D C B C B

Translation

1 在每次飞行中，都有许多看不见的力在起作用。

2 我是一名退休的航空工程师，在航空领域工作了36年后，从美国航空航天局兰利研究中心退休。

3 空气阻力并不是拉力，而是阻止飞机向前运动的力。

4 我的工作就是找出降低音爆的方法，以便减轻由于超音速飞行而产生的音爆的干扰。

5 气球破裂产生的高压冲击波，呈球体状向四面八方蔓延开来。

6 它就像一个以机头为顶点的圆锥体（类似冰激凌筒），所有受到飞机干扰的分子会被推到锥体外部，较高压力的空气会位于锥体内部，而这个圆锥体会一直发散延伸到地面。

7 （在风洞中）空气实际上是吹向飞机模型的，这和飞机实际飞行时所产生的气动效果几乎相同，而我们就在风洞中测量想要的数据点。

8 兰利全尺寸风洞是最著名的风洞之一，它是一个与足球场一样长的室内风洞（封闭在建筑物内），其有一个测试区，用于测试全尺寸飞机。

9 我们希望能把音爆降低到不会对人造成干扰的程度。

10 我们的最终目标就是利用科学原理和科技进步来造福社会和人民。

Expansion Exercise for CET-BAND 4—Reading

1~5 A B C B C

Unit 3

Listening

1~5 B G E F H

6~10 A D J C I

Reading Exercises

I. Vocabulary Matching

1~5 O B C D N

6~10 F G H I J

11~15 K L M E A

16~20 P Q R X U

21~25 T V W S Y

II. Answer the Following Questions

1 Motivation acts as the engine that propels the organization towards its objectives. It encourages employees to excel in their roles and contribute their best efforts and ideas for

success.

2　Effective communication is vital for seamless coordination, information sharing, and collaboration among team members. It helps prevent misunderstandings, ensures alignment, and maximizes opportunities for success.

3　Just as smooth coordination is crucial for a successful flight, effective teamwork, collaboration, and coordination are essential for achieving collective goals in organizations. Lack of coordination can lead to turbulence and hinder progress.

4　Human factors trainings equip employees with the necessary knowledge and skills to handle unexpected and complex situations. They prepare individuals to operate effectively under maximum stress and help organizations navigate through challenging times.

5　Organizations need to stay vigilant and monitor external threats to anticipate and mitigate potential obstacles. By staying informed about industry trends, competitors, and emerging risks, organizations can make proactive decisions and adapt their strategies accordingly.

III. Multiple Choice

1~5　C A C D C

Translation

1　我们开发了可以迅速、有效、安全地克服困难的各种工具。

2　通过窗户，您可以看到山峦逐渐消失，仿佛您在行业中面临的各种阻力烟消云散——如财务压力、增长目标、自动化程度提高、行业竞争或员工短缺等。

3　企业如同飞机，其"机翼"就是员工的挑选、教育及培训，但仅凭它无法使其起飞。

4　只要得到良好保养和维护，发动机的性能就会极其稳定。

5　事实就是：如果没有燃料，什么都无法运作。

6　即便没有外部因素的影响，控制和协调的缺失也可能会导致飞行颠簸。

7　优秀的飞行员关注雷达，并提前做好预判和躲避风险的准备。

8　在公司内部，当面临的压力达到顶峰，最关键的战略储备就是所有必要的培训，它们能够帮助公司应对意外和复杂情况。

9　飞机装配了合适的应急设备、逃生出口和滑梯，以便每名乘客都能够安全迅速地撤离。

10　组织及个人，特别是其间的高效协作发挥着至关重要的作用，他们决定成败，今日和未来皆是如此。

Expansion Exercise for CET-BAND 4—Reading

1~5　B D A D A

Unit 4

Listening
1~5　J C B F I
6~10　A G H D E

Reading Exercises
I. Vocabulary Matching
1~5　G B C R E
6~10　F A H I V
11~15　K L M N O
16~20　P Q D S T
21~22　U J

II. Answer the Following Questions
1　Aviation offers various roles in HR, IT, sales, project management, aircraft maintenance, airline administration, airline law, airline management, passenger assistance, avionics technology, crew scheduling, flight instruction, and human resources.
2　Aircraft maintenance technicians usually have a high school diploma and are responsible for inspecting, maintaining, and repairing aircraft components and systems.
3　Airline pilots are responsible for safely transporting passengers and cargo, following safety procedures, navigating through different weather conditions, and communicating with air traffic control and cabin crew.
4　Airline ramp agents ensure passenger and cargo safety, handle baggage and cargo loading/unloading, and provide ground guidance for aircrafts.
5　Flight attendants need strong customer service skills, stress management, and cabin safety oversight. Salaries vary based on experience, with entry-level around $35,000 per year and senior-level up to $75,000 per year. Benefits may include insurance, vacation, 401(k) plans, and some airlines offer sign-on bonuses and incentives.

III. Multiple Choice
1~5　C B C B A

Translation
1　说起航空职业，有句老话说："天空就是我家。"这句话形容的不仅仅是飞行人员。
2　机务（飞机维修技术员）负责飞机各部件和系统的检查、维护和维修。
3　如果你对航空职业感兴趣，那么航空公司行政管理岗位也许适合你。
4　航空公司法务必须拥有正规法学院的法律学位，在航空公司所在州获得律师执业执照，并有几年的航空法务经验。

5　飞行员必须确保执行所有的安全程序，同时他们也必须能够处理可能发生的紧急情况。

6　航空公司销售代表也要处理客户投诉，提供必要的帮助。

7　航空电子设备技术员负责维护和安装飞机上的通讯、导航和监控设备。

8　要成为一名成功的空乘人员，拥有强大的客户服务技能和有效处理压力的能力是很重要的。

9　空乘人员在确保飞机上乘客的安全方面发挥着至关重要的作用。

10　飞行教员通常是从具有较强飞行素质和多年飞行经验的资深飞行员中招募的。

11　除非他们穿着某一航空公司的制服，否则你看到的所有机场工作人员都受雇于机场，或是受雇于在机场有特许经营权的公司或承包商。

12　机场车辆机械师也是进入航空行业的一个很好的选择，因为凭借经验，他们可以晋升到更专业的岗位。

13　飞行调度员的角色对乘客和机组人员的安全至关重要。他们负责协调与飞行有关的所有地面工作，包括装卸行李和货物、为飞机加油和安排餐饮服务。

Expansion Exercise for CET-BAND 4—Reading

1~5　E M D I H

6~10　K F L B A

Unit 5

Listening

1~5　A E F H B

6~10　I G C J D

Reading Exercises

I. Vocabulary Matching

1~5　T S R Q P

6~10　O N M L K

11~15　J I H G F

16~20　E D C B A

II. Answer the Following Questions

1　Feng Ru was an engineer and inventor who made significant contributions to aviation. He established an aircraft factory and built airplanes of his own design.

2　Feng Ru tested his airplane above the hills of Oakland, California.

3　Feng Ru returned to California to establish an aircraft factory after being forced to relocate from San Francisco due to the earthquake and fire.

4　Feng Ru had the engine castings made by different machine shops and assembled the parts himself to maintain secrecy.

5　Feng Ru received a full military funeral and was posthumously awarded the rank of a major general by the Republic of China. The words "Chinese Aviation Pioneer" were engraved on his tombstone.

III. Multiple Choice

1~5　A B C A C

Translation

1　冯如被美国的国力和繁荣震撼了。他认识到，是工业化使美国强大；他也意识到工业化也可以使中国变得同样富强。

2　但在听说赖特兄弟的成功后，冯如把注意力转向了航空，不辞辛苦地把他找到的任何关于赖特兄弟、格伦·柯蒂斯以及后来的法国飞机设计师亨利·法曼的相关资料都翻译成了中文。

3　到1906年，冯如决定回到加利福尼亚建立一个飞机工厂，制造他自己设计的飞机。

4　经美国航空航天学会证实，在1909年，在冯如的小小工坊，这位自学成才的工程师成立了广东飞行器公司，并在当年完成了他的第一架飞机的制造。

5　冯如的谨慎最终得到了回报：他的成功试飞被主流媒体报道，他的工作得到了革命家孙中山先生的赞扬。

6　据报道，冯如在弥留之际，勉励助手："勿因吾毙而阻其进取之心，须知此为必有之阶段。"

Expansion Exercise for CET-BAND 4—Reading

1~5　E C L H J

6~10　B M D A I

Unit 6

Listening

1~5　G F C A J

6~10　H I D E B

Reading Exercises

I. Vocabulary Matching

1~5　D A B M N

6~10　S I E T G

11~15　H O U L R

16~20　F Q J K P

21　C

II. Answer the Following Questions

1　The pilots of US Airways flight 1549 were Captain Chelsey Sullenberger, also known as "Sully", and First Officer Jeffrey Skiles.

2　The emergency situation was caused by a collision with a flock of Canada geese, which resulted in both engines of the aircraft shutting down.

3　Captain Sullenberger took control of the aircraft while the co-pilot attempted to restart the engines. When the engines couldn't be restarted, the co-pilot turned on the auxiliary power unit (APU) to maintain some control over the plane. They initiated a gliding descent towards the Hudson River.

4　Captain Sullenberger realized that the plane wouldn't be able to make it back to the airport safely, so he made the decision to land in the Hudson River. He believed it was the only suitable location in the densely populated area where the plane could land safely.

5　All passengers and crew members survived the emergency landing, with only five serious injuries and 78 people receiving treatment for minor injuries and hypothermia. The successful landing and subsequent rescue operation were recognized and awarded for their remarkable achievements.

III. Multiple Choice

1~5　A C D B A

Translation

1　空中客车 A-320 的机长是 57 岁的切尔西·萨伦伯格（绰号"萨利"），他是航空安全方面的专家，曾是一名滑翔机飞行员。

2　在这个仿佛命中注定的日子，萨伦伯格已经累积了超过 19000 小时的飞行时间，其中几乎有 5000 小时是在 A320 上。

3　驾驶舱机组在这段时间里，将飞机不断升高并执行了一系列必要的飞行任务。

4　在海拔 2800 英尺（约 854 米）的高空以如此高的速度移动，飞行员无法避免碰撞。

5　然后，飞机机体震颤，机内的每个人都听到了引擎传来的隆隆声和震耳欲聋的轰鸣声。

6　However, Sully didn't lose his presence of mind, even after he realized that both engines had shut down.

7　With the APU on, the plane was still climbing, but it was visibly slowing down.

8　The plane flew less than 900 ft (270 m) over the George Washington Bridge while the co-pilot was calling out the airspeed and altitude of the plane.

9 The cockpit was filled with warnings repeated by a computerized voice, "Terrain, too low, too low, caution terrain, pull up, pull up, pull up".

10 But naturally, that wasn't the end of the story—the crew had to evacuate people from a plane that was floating in the middle of the river.

Expansion Exercise for CET-BAND 4—News Report
1~2 B C

Unit 7

Listening
1~5 B H G J C
6~10 E A D F L

Reading Exercises
I. Vocabulary Matching
1~5 B C J D E
6~10 A H F K L
11~15 M O P Q G

II. Answer the Following Questions
1 Some benefits of utilizing AI in aviation include streamlining routes, reducing emissions, enhancing the customer experience, and optimizing mission outcomes.

2 One significant challenge is the absence of a standardized definition for AI due to its constant evolution. AI's complexity surpasses traditional algorithms, as it learns from experience and adapts based on new data. Achieving mandatory certification determinism for evolving AI programs also poses a challenge.

3 Safety concerns revolve around the trustworthiness of AI and potential errors or safety violations. Questions arise about AI's ability to emulate human performance and the need for deterministic AI learning and real-time monitoring. Effectively communicating AI operations to stakeholders is also a challenge.

4 Regulatory bodies such as the Federal Aviation Administration and the European Union Aviation Safety Agency (EASA) are actively interested in AI in aviation. EASA published a report advocating for a human-centric approach to AI programs. Manufacturers like Boeing and Airbus are investing in AI research, and industry organizations like SAE are developing aviation standards and training materials.

5 Alaska Airlines conducted a trial of an AI-driven program called "Flyways" during the pandemic slowdown. Flyways optimized flight paths based on factors like the original route,

current weather conditions, and aircraft weight. Real-time adjustments led to an average time reduction of five minutes per flight, resulting in significant fuel savings of 480 thousand gallons during the trial period.

III. Multiple Choice
1~5　C B A B C

Translation
1　阿拉斯加航空公司利用 AI 来优化飞行路径，实现节省燃料和碳中和的目标。
2　FAA 和 EASA 是监管机构，它们关注 AI 在航空领域的发展，并致力于提高其可信性，坚持以人为本的理念。
3　在航空领域，定义 AI 存在挑战，因为 AI 不断演进的特性：它允许计算机通过自身经验自主学习，并会根据新数据调整响应。
4　阿拉斯加航空公司通过名为 Flyways 的 AI 驱动程序来优化航线、测试不同的航线、收集数据，并根据实时情况进行调整，以创建高效的飞行路径。
5　航空业中 AI 的未来前景广阔，行业正不断投资研究和技术发展，预计 AI 将逐步进入驾驶舱，类似于其在汽车行业的发展路径。
6　FAA 和 EASA 正在探索 AI 的安全问题，包括其定义、安全性以及在航空领域中对 AI 操作的明确交流问题。
7　根据航空安全机构 AFuzion 的说法，预期在 2030 年左右会出现重要的驾驶舱 AI 解决方案。

Expansion Exercise for CET-BAND 4—News Report
1~5　D A C

Unit 8

Listening
1~5　F C B G E
6~10　I H J D A

Reading Exercises
I. Vocabulary Matching
1~5　P Q E I J
6~10　K A M B C
11~15　D G R V H
16~20　T S U N O
21~23　W L F

II. Answer the Following Questions

1 Justin became interested in becoming a pilot when he visited the cockpit of a Delta Air Lines flight as a child and was fascinated by the experience.

2 Justin worked as a ramp agent for Delta Air Lines and received guidance from a pilot mentor. He attended flight school, became a flight instructor, and eventually got a job as a pilot at Breeze Airways.

3 Justin posted a tweet with before-and-after photos of his journey, and it gained attention and went viral on Twitter, with many people inspired by his story.

4 Justin encouraged donations to the Organization of Black Aerospace Professionals, which he mentioned in his tweet.

5 Delta Air Lines is actively working to create a diverse pilot workforce by removing barriers and expanding recruitment efforts. They have programs like Propel that help employees from various roles within Delta progress towards becoming pilots.

III. Multiple Choice

1~5 D A C B C

Translation

1 本公司采用名为 Propel's Company Pathway 的职业发展项目，目前有 100 多名员工参与其中各阶段，包括在达美航空担任过空乘、机坪工作人员、登机口工作人员、机械师、飞行调度员和其他行政或后勤岗位的人。

2 在一辆机场员工巴士上，他与当时是维珍美国航空公司飞行员的伊沃尔·马丁交谈，提到他虽然喜欢做机坪工作人员的工作，但他真正想成为的是一名飞行员。

3 这次简短的交流发展成了一段导师关系，因为马丁帮助穆塔瓦西姆规划了实现梦想所需的实际步骤。

4 达美航空致力于组建多元化的员工队伍，以回馈我们在国内和世界其他地区服务的社区，这一承诺坚定不移。

Expansion Exercise for CET-BAND 4—Reading

1~5 F J B I M

6~10 C G A K E

Unit 9

Listening

1~5 A I C B E

6~10 F D G H J

Reading Exercises

I. Vocabulary Matching

1~5　D I F B E

6~10　Q G H C J

11~15　R K P N O

16~20　L S A M V

21~24　U X W Z

II. Answer the Following Questions

1　Airlines overbook to make more money and use their resources efficiently. They sell extra tickets because they expect some passengers to not show up for their flights.

2　Airlines use past data on customer no-show rates to figure out how many tickets to sell. They analyze the probabilities of passengers showing up and use methods like the binomial distribution to estimate the right number of tickets.

3　Revenue is important for airlines when deciding whether to overbook. They consider the income from ticket sales and compare it to the costs of bumping passengers. By calculating different scenarios, they aim to find the balance that brings in the most money.

4　Airlines use the binomial distribution to calculate the chances of different passenger counts. This helps them predict the probabilities of various outcomes based on past data. By looking at these probabilities, airlines can estimate the expected number of passengers and its impact on revenue.

5　The ethics of overbooking are debated. Some argue it's wrong to sell more tickets than seats available, while others say it can be acceptable if there's a high chance of no-shows. The ethical threshold between practicality and customer satisfaction varies, and different people have different views on what's ethically acceptable.

III. Multiple Choice

1~5　A A C A B

Translation

1　在美国，每年约有 5 万人的舱位由于航空公司超售机票而被取消。

2　航空公司售票赚钱，但一旦乘客被取消舱位就会使公司赔钱。

3　带入不同的额外票数来重复这种计算，航空公司可找到此例中能产生最高收入的票数。

4　如果不超售机票，你只能获利 45000 美元；而如果超售了 15 张机票，并且至少有 15 人未登机，你将获利 48750 美元。

5　我们可以通过在这个例子中使用二项分布来找出，195 名乘客全部登机的概率几乎为零。

6 While often infuriating for the customer, overbooking happens because it increases profits while also letting businesses optimize their resources.

Expansion Exercise for CET-BAND 4—Reading

1~5　C B A C D

Unit 10

Listening

1~5　C D E H I

6~10　J A B F G

Reading Exercises

I. Vocabulary Matching

1~5　A D F C K

6~10　G M J T E

11~15　U P N O B

16~20　I R S L Z

21~25　H W U V Y

26　Q

II. Answer the Following Questions

1 China's development of the C919 passenger jet began in 2008.

2 COMAC faced challenges such as technical complexities, supply issues, and restrictions on component shipments imposed by the Trump administration.

3 The first C919 prototype was published in 2015.

4 The C919 aims to compete with planes like the Boeing 737 Max and Airbus A320 by offering lower prices and becoming a strong rival in the passenger jet market.

5 China's goal is to produce all the components of the C919 domestically, reducing its reliance on foreign technology and becoming self-sufficient in the manufacturing process.

III. Multiple Choice

1~5　C D C D A

Translation

1 尽管面临技术复杂性和供应问题，中国商用飞机有限公司仍致力于提供性能优秀的飞机。

2 C919项目始于2008年，在此过程中遇到了许多挑战，项目主创人员坚持不懈并取得了显著进展。

3 中国的雄心不仅限于生产自己的客机，她旨在实现C919所有零部件的制造自给自足。

4 C919的首航是中国航空业的重要里程碑，展示了其技术进步和在全球舞台上竞争的决心。

5 中国追求生产和出口自己的客机，对于老牌制造商如波音和空客的主导地位构成了潜在挑战。

6 中国商用飞机有限责任公司与庞巴迪航空等国际公司的合作有助于提升C919的能力和整体质量。

7 C919的成功研发和商业化有可能打破波音和空客在全球航空市场上的双寡头垄断局面。

8 C919具有竞争力的定价、先进的技术和可靠的性能使其在全球航空市场上成为了一个强大的竞争对手。

Expansion Exercise for CET-BAND 4—Translation

The Grand Canal is the longest man-made river which extends all the way from Bejing in the north to Hangzhou in the south, and one of the grandest projects in the Chinese history. The construction of the canal was started in the 4th century BC and was completed in the 13th century. It was first built for grain transportation and later also for transporting other commodities. The areas along the canal have gradually developed into the industrial and commercial centres of China. For a long period of time, the canal has been playing a significant role in the development of the Chinese economy, greatly enhancing the personnel exchange and cultural communication between northern and southern regions.

Unit 11

Listening

1~5　H B E I C

6~10　A J D G F

Reading Exercises

I. Vocabulary Matching

1~5　G F E J O

6~10　M A C I L

II. Answer the Following Questions

1　In 2019, direct CO_2 emissions from aviation accounted for at least 2% of global emissions.

2　Some proposed alternative technologies include zeppelins, blended-wing aircraft, solar power, and biofuels made from sustainable sources.

3 KLM faced legal action for misleading marketing claims regarding its biofuel usage.

4 One notable quality of hydrogen is its potential for environmentally friendly production using renewable energy (green hydrogen).

5 A combination of technological innovations, changes in consumer behavior, and sustainable practices such as improving fuel efficiency, optimizing flight routes, and offsetting emissions through carbon offset programs is necessary to achieve a more eco-friendly future of aviation.

III. Multiple Choice

1~5 B D B C D

Translation

1 航空业已成为气候危机的主要原因，其碳足迹构成了重大威胁。

2 2019年，航空产业的直接二氧化碳排放量至少占全球排放量的2%，使其成为最大的能源密集型人类活动。

3 然而，随着市场放松管制，低成本航空公司应运而生，以满足日益增长的航空旅行需求。航空从少数人的特权变成了服务许多人的日常交通方式。

4 值得庆幸的是，航空业多年来提出了许多替代技术，为更环保的未来增添了希望。

5 特别是生物燃料，已经展示出减少二氧化碳排放的前景，但其有限的可用性仍然是一个挑战。

6 氢具有三个显著的特性：作为能量载体的多功能性，可以为高动力需求的飞机提供动力；作为可再生能源进行环保生产的潜力（绿氢）；在燃料电池中使用时只排放水蒸气。

7 虽然氢燃料很有发展的希望，但目前它并不是一个放之四海而皆准的解决方案。

8 最终，包括技术创新和消费者行为改变在内的多种方法的结合将是实现更环保的航空未来所必需的途径。

Expansion Exercise for CET-BAND 4—News Report

1~2 C A

Unit 12

Listening

1~5 E F G I D

6~10 A H C J B

Reading Exercises

I. Vocabulary Matching

1~5　I B C T E
6~10　F S H A J
11~15　K L U N O
16~20　Q P R G D
21~22　M V

II. Answer the Following Questions

1　The pilots relied on their visual cues and used a glass of water placed on the control panel to measure the angle of descent.

2　The landing strip near Izhma was abandoned and shorter than the recommended length for a safe landing. The aircraft's speed during descent exceeded the safe threshold, and the flaps could only be operated via electrical switches.

3　The pilots made three attempts, carefully evaluating the risks and adjusting their approach. They brought the aircraft down onto the runway, applied maximum reverse thrust, and engaged the emergency brakes to decelerate and bring the plane to a stop just meters from the end of the runway.

4　The passengers remained remarkably calm and composed throughout the ordeal. They understood the gravity of the situation and patiently awaited their fate, ultimately expressing relief, disbelief, and gratitude when the aircraft came to a halt.

5　The investigation revealed that a faulty component in the electrical system was the cause of the equipment malfunction. The component had gradually deteriorated over time, leading to the failure of the electronic systems on board the aircraft.

III. Multiple Choice

1~5　D C A B B

Translation

1　乘客们上了飞机，空乘员漫不经心地核对了他们的登机牌，一切都如往常一样。

2　机长叶甫盖尼·诺沃索洛夫向乘客保证：飞机正在经历轻微的湍流，飞行可能会在短时间内变得有些颠簸。

3　它仍然可以运行，但所有导航系统都无法工作，无线电中断，燃油传输泵也丧失了功能。

4　即使在其他条件完美无缺的情况下，飞行时间也不足以到达最近的机场，但是所有电子设备都无法工作，要找到一个合适的着陆点并且在那里安全着陆，实际上是几乎不可能的。

5　Taking turns, the pilots watched the water in the glass and measured the angle of descent by its incline.

6 Despite the abandonment of the airport, Sotnikov, of his own will, kept the runway in good condition for all those years.

7 However, the length of the runway in Izhma was only 4,347 ft (1,325 m), while this airplane needs 6,560 ft (2,000 m) to land safely.

8 So, against all odds, with none of the plane's electrical systems working, with no proper runway to land on, with a speed that should have crashed the aircraft into the forest, and without even the means to tell anyone they were having an emergency, the heroic pilots still landed the plane and saved everyone's lives.

9 Their unwavering determination to save lives in the face of overwhelming odds serves as an inspiration to aviation professionals and a reminder of the importance of human expertise in the age of advanced technology.

Expansion Exercise for CET-BAND 4—News Report

1~3 B D B

References

1. Cleave, K. V. How China Developed Its First Large Domestic Airliner to Take on Boeing and Airbus. https://fortune.com
2. Flight Jets World. How Genius Pilots Used a Glass of Water to Save a Plane From Crashing. https://fighterjetsworld.com/
3. Gagnon, D. The Sky Is the Limit: Types of Airport Jobs and Aviation Careers. https://www.snhu.edu
4. Glenn Research Center. Four Forces on an Airplane. https://www1.grc.nasa.gov
5. Hayward, J. The Miracle On The Hudson. https://simpleflying.com
6. Hilderman, V. AI in the Sky: How Artificial Intelligence and Aviation Are Working Together. https://interactive.aviationtoday.com
7. HyScaler Inc. Elevating Airforce Excellence: The Future Flight with AI Wings and Cutting-edge tools. https://hyscaler.com
8. Indeed. 18 Jobs in the Aviation Industry. https://www.indeed.com
9. Klietsch, N. Why Do Airlines Sell Too Many Tickets? https://www.ted.com
10. Lewis, M. The Importance of Human Facto Training. https://www.missionperformance.com
11. Maksel, R. The Father of Chinese Aviation. https://www.smithsonianmag.com/
12. Murrhy, B. A Delta Air Lines Pilot's Inspiring Tweet Went Viral. The Reason Why Is Brilliant. https://www.inc.com
13. Prisco, J. Will Guilt-free Long-haul Flight Ever Be Possible? https://edition.cnn.com
14. Rafferty, J. P. Why Do Airlines Overbook Seats on Flights? https://www.britannica.com/
15. The Wall Street Journal. How A Digital Health Passport Could Work. https://www.wsj.com
16. Tikkanen, A. US Airways Flight 1549. https://www.britannica.com/
17. Usman, R. Human Factors in Safety: Types, Examples, and Solutions. https://www.hseblog.com/
18. Virtual Jet Centre. Human Factors. https://www.virtualjetcentre.co.uk
19. Wang, Y. C919 Jetliner Complete Debut Commercial Flight. https://global.chinadaily.com.cn
20. World Economic Forum. Meet the World's First Zero-emission, Hydrogen-powered

Aeroplane Engine. https://www.weforum.org

21. World Economic Forum. Could hydrogen-fuelled flights be a reality by 2035? https://www.weforum.org
22. Wrigley, S. The "Lucky" Landing of the Tupelov Tu-154. https://fearoflanding.com
23. https://e-aircraftsupply.com/the-science-behind-aircraft-flight/
24. https://en.wikipedia.org/wiki/Feng_Ru
25. https://en.wikipedia.org/wiki/US_Airways_Flight_1549
26. https://iata.org/en/careers/
27. https://medium.com/@siphokhumalo/skybound-intelligence-how-artificial-intelligence-is-revolutionizing-the-future-of-aviation-faa33e501278
28. https://weforum.org/agenda/2022/07/how-to-achieve-net-zero-in-aviation/
29. https://interactive.aviationtoday.com/avionicsmagazine/may-june-2022/ai-in-the-sky-how-artificial-intelligence-and-aviation-are-working-together/